Advances in Microporous and Mesoporous Materials

Edited by Rafael Huirache Acuña

Published in London, United Kingdom

IntechOpen

Supporting open minds since 2005

Advances in Microporous and Mesoporous Materials
http://dx.doi.org/10.5772/intechopen.83325
Edited by Rafael Huirache Acuña

Contributors
Anjali Patel, Priyanka Solanki, Rafael Huirache Acuña, Cristina Farias Rosales, Rut Guil López, María Faraldos, Rafael Maya Yescas, Trino Zepeda, Barbara Pawelec, Leena Nebhani, Smrutirekha Mishra, Tina Joshi, Minghua Wang, Soledad Aspromonte, Federico Piovano, Alicia Boix, Esther Alonso, Hui Zhao, Gulambar Tursun, Jianwang Yang, Long Liu

First published in London, United Kingdom, 2020 by IntechOpen
IntechOpen is the global imprint of INTECHOPEN LIMITED, registered in England and Wales, registration number: 11086078, 7th floor, 10 Lower Thames Street, London, EC3R 6AF, United Kingdom
Printed in Croatia

British Library Cataloguing-in-Publication Data
A catalogue record for this book is available from the British Library

Additional hard and PDF copies can be obtained from orders@intechopen.com

Advances in Microporous and Mesoporous Materials
Edited by Rafael Huirache Acuña
p. cm.
Print ISBN 978-1-83880-755-9
Online ISBN 978-1-83880-756-6
eBook (PDF) ISBN 978-1-83880-757-3

For EU product safety concerns:
IN TECH d.o.o., Prolaz Marije Krucifikse Kozulić 3, 51000 Rijeka, Croatia, info@intechopen.com or visit our website at intechopen.com.

Meet the editor

Dr. Rafael Huirache Acuña studied Chemical Engineering (1995–2000) and attained his master's degree in Metallurgy and Materials Science from the Institute of Metallurgical Research-UMSNH, Mexico, in 2003. He obtained his PhD in 2007 in Materials Science at the Advanced Materials Research Center (CIMAV, SC) working on the synthesis, characterization, and testing of trimetallic catalysts for HDS reactions, with a stay in ICCOM Florence. After his PhD, he started postdoctoral research devoted to mesoporous supports for HDS catalysts at the Center of Applied Physics and Advanced Technology of UNAM (2008–2009). In 2010 he reached the staff position of professor and researcher at the Chemical Engineering Faculty of UMSNH, Mexico. His research interests are focused mainly on heterogeneous catalysts and materials for environmental applications.

Contents

Preface

Nowadays, microporous and mesoporous materials are versatile solids of great interest due to their structures and pore sizes, which allow their application in many areas. Accordingly, with IUPAC, materials with pore sizes smaller than 2 nm are called microporous and with pore sizes between 2 and 50 nm are called mesoporous. The pore size has an important impact on the material properties and affects their applications. In addition, high surface area, and their ability to incorporate functional groups on the framework are of great relevance for commercial and science applications. This book intends to provide readers with a comprehensive overview of recent improvements in the microporous and mesoporous materials field.

Advances in Microporous and Mesoporous Materials focuses on important research presented in the following chapters: "Synthesis of supported mesoporous catalysts using supercritical CO_2," "Sulfided NiMo/*clinoptilolite* catalysts for selective sulfur removal from naphtha stream without olefin hydrogenation," "Functionalized MCM-48 as carrier for in vitro controlled release of an active biomolecule, L-arginine," "Polymer functionalization of mesoporous silica nanoparticles using controlled radical polymerization techniques," and "Melting and dissolving fly ash by NaOH for the removal of iron, calcium and other impurities."

I believe that the information included in this book shows interesting and various advances in the microporous and mesoporous materials field.

I wish to acknowledge Ms. Mia Vulovic, Author Service Manager, and Ms. Kristina Jurdana, Commissioning Editor, for their important support during this project. I would also like to thank to IntechOpen for their invitation to participate as academic editor.

Dr. Rafael Huirache Acuña
Professor and Researcher,
Universidad Michoacana de San Nicolás de Hidalgo,
Morelia, Michoacán, México

Chapter 1

Synthesis of Supported Mesoporous Catalysts Using Supercritical CO_2

Soledad Guadalupe Aspromonte, Federico Andrés Piovano, Esther Alonso and Alicia Viviana Boix

Abstract

Metal and metal oxide nanoparticles have attracted increased attention due to their unusual physical and chemical properties. The nature, dispersion, and size of the nanoparticles are key factors in determining the activity and selectivity of the supported catalysts. Supercritical fluid deposition (SCFD) is a promising method to deposit metallic nanoparticles and films on inorganic porous supports. CO_2 is the most commonly used supercritical fluid (sc-CO_2) for material synthesis because it is nontoxic, nonreactive, nonflammable, and inexpensive. This work presents the synthesis of cobalt, nickel, and ruthenium nanoparticles on MCM-41, Al-MCM-41, MCM-48, and activated carbon supports in sc-CO_2. Batch and continuous deposition are studied, with two high-pressure reactor configurations: column or alternative (sandwich). To avoid the length of the bed being too long, the reagents were separated into smaller amounts and placed alternately, keeping the total mass of the precursor and support constant. The prepared samples were characterized by scanning electron (SEM/EDX) and transmission electron microscopy (TEM).

Keywords: metal, ruthenium, cobalt, nickel, supercritical fluids

1. Introduction

The chemical industry is partially responsible for the environmental problems caused by many of its processes, such as air pollution due to emissions produced by the incineration of fossil fuels. Thus, Green Chemistry emerges in response to these issues [1, 2]. This represents a constant motivation for the design of new processes that minimize the use of dangerous substances and the generation of polluting agents, maintaining the feature of the final product.

In this context, industrial processes should be designed to generate substances that are slight or not harmful to human health and the environment and, if it is possible, to use auxiliary agents (e.g., solvents) that are innocuous or with low toxicity. According to this dogma, in recent years, several investigations have been carried out involving the replacement of solvents used in many industries by other reaction media such as ionic liquids [3–5] or supercritical fluids (SCFs) [6–8].

Micro- and mesoporous materials are solid materials widely used in heterogeneous catalysis, due to their involvement in numerous reactions of industrial interest.

The deposition of metallic active phases on these substrates is of utmost importance in order to maintain the textural and structural properties and achieve good dispersion.

In this chapter, the study of supercritical carbon dioxide (sc-CO_2) as a solvent to deposit metallic nanoparticles (Ni, Co, Ru) on microporous (activated carbon) and mesoporous substrates (MCM-41 and MCM-48) is presented.

Supercritical fluid technology satisfies the principles of Green Chemistry, because it is a reagent with low toxicity and whose critical parameters are easily accessible and, therefore, energy consumption is minimized. In addition, through its compression it can be recovered, thus avoiding its release into the atmosphere. The combination of all these aspects, together with its good transport properties, as will be discussed in detail below, makes sc-CO_2 an excellent resource for developing new environmentally friendly technologies.

2. Supercritical fluids (SCFs)

A supercritical fluid (SCF) is defined as the state of a substance whose pressure and temperature are higher than their critical values [9]. In this state, the phase boundary between the liquid and the gas is interrupted at the critical point, which implies the formation of a homogeneous phase as observed in **Figure 1**. Therefore, it is possible for a substance to cross from the liquid to gaseous state passing through the supercritical region, without any phase transition (**Figure 2**).

Supercritical fluids have properties that are in many aspects unique and change considerably from those corresponding to the liquid or gaseous state of origin. Basically, SCFs have intermediate properties between those that characterize a liquid and those of a gas [10]. The thermal conductivity of SCFs approximates that of liquids, indicating that they have better properties with respect to heat conduction [11]. SCFs tend to occupy the entire volume of the enclosure and do not present interfacial tension, a behavior similar to gases (**Table 1**). Similar to gases, the density of SCFs varies enormously with pressure and temperature, although densities very close to those of liquids are reached. Thus, the characteristic property of SCFs is the wide range of high densities that they can adopt depending on pressure and/or temperature conditions. This differentiates them from liquids that are practically incompressible and from gases that have very low densities. These properties make FSC an effective medium for incorporating metals into porous solids.

Figure 3 presents the isotherms for a fluid, indicating the variation in density with pressure. It is noted that, above the critical point, there is only one phase and for high densities, the curve remains flat, indicating large variations in density for small increases in pressure.

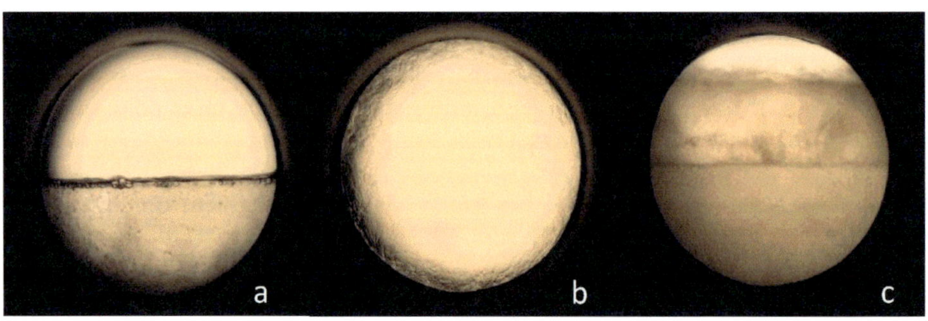

Figure 1.
Phases observed with a high pressure cell: (a) heterogeneous gas/liquid, (b) phase transition and (c) homogeneous in supercritical state.

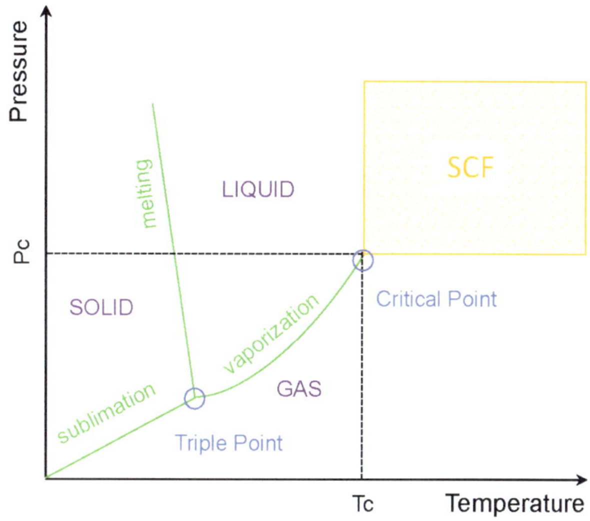

Figure 2.
P-T phase diagram.

Properties	Gas[a]	SCF[b]	Liquids[a]
Density ($kg{\cdot}m^{-3}$)	0.5–2.0	200–500	500–1500
Viscosity (mPa·s)	0.01–0.30	0.01–0.03	0.2–3.0
Diffusivity ($m^2{\cdot}s^{-1}$)	10–5	10–7	10–9
Conductivity ($W{\cdot}m^{-1}{\cdot}K^{-1}$)	0.01–0.02	0.05–1.00	0.1–0.2

[a]*Properties at room temperatures.*
[b]*Properties close to critical point.*

Table 1.
Typical values for supercritical fluids (SCFs) [10].

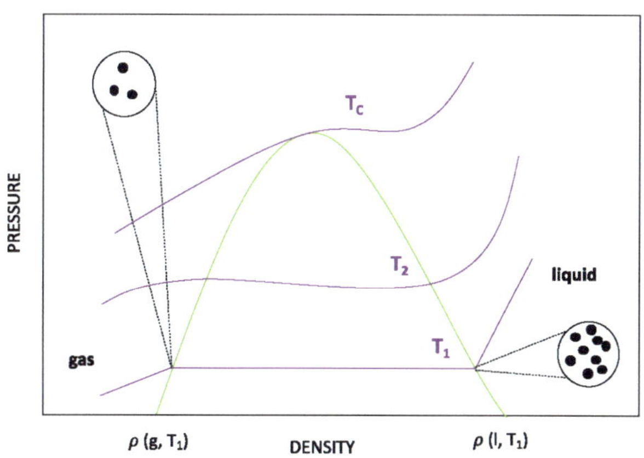

Figure 3.
Pressure-density diagram.

Therefore, considering the direct ratio that exists between fluid density and solvation power, SCFs can greatly vary their solvation capacity by small variations in pressure and/or temperature. Consequently, knowledge of the thermodynamic properties of SCFs and their behavior near the critical point is essential for the proper design of separation reactions and processes.

Thus, the main applications of supercritical fluids are [12–14]:

- *Extraction*: leaves no residue; high purity extracts are obtained and do not require high temperatures.
- *Synthesis of materials*: SCFs allow obtaining solid materials with controlled properties, among which the obtaining of aerogels and the synthesis of ultra-fine particles (nanoparticles) with very uniform morphology, high purity, and free of solvent residues can be highlighted.
- *Reaction medium*: the existence of a single phase allows optimal mass and energy transfer.

Indisputably, the most used supercritical fluid in research and industrial applications is carbon dioxide (CO_2). It is an innocuous, abundant, and inexpensive gas, with relatively low critical conditions (31°C and 71 bar) and therefore easy to operate.

Table 2 presents the temperature, pressure, and density conditions for different supercritical fluids [15].

2.1 Synthesis of materials in supercritical fluids

Currently, the material synthesis using SCFs represents a new area of research, particularly, in the field of nanoparticles. Polymeric materials, inorganic solids, and organic compounds, with uses in additives, medicines,

SCF	Name	T_c [a]	P_c [b]	ρ_c [c]
C_2H_4	Ethylene	9.3	50.4	220
Xe	Xenon	16.6	58.4	120
CO_2	Carbon dioxide	31.1	73.8	470
C_2H_6	Ethane	32.2	48.8	200
N_2O	Nitrous oxide	36.5	71.7	450
C_3H_8	Propane	96.7	42.5	220
NH_3	Ammonia	132.5	112.8	240
C_3H_8O	1-Propanol	235.2	47.6	270
CH_4O	Methanol	239.5	81.0	270
H_2O	Water	374.2	220.5	320
C_7H_8	Toluene	318.6	42.1	290
C_4H_{10}	n-Butane	152.0	38.0	228
C_5H_{12}	n-Pentane	196.6	33.7	232
C_6H_6	Benzene	289.5	49.2	300

[a] *Critical temperature (°C).*
[b] *Critical pressure (bar).*
[c] *Critical density (kg m^{-3}).*

Table 2.
Critical properties of different fluids.

pigments, cosmetics, biomaterials, optics, catalysts, or adsorbents, have been synthesized with supercritical fluids [16].

As a consequence, the development of various methods on the generation of particles with supercritical carbon dioxide (sc-CO_2) has grown rapidly in the last decades [17]. Among them, the rapid expansion of supercritical solutions (REES) is one of the most studied and used precipitation processes on an industrial and/or laboratory scale [18, 19]. This method involves the rapid expansion of a pressurized solution through a needle, causing extremely rapid nucleation of the solid solute in micronized form [20]. However, the rapid expansion of a supercritical solution in a liquid solvent (RESOLV) is an alternative to the REES technique [21]. In this case, expansion occurs in a liquid solvent instead of using sc-CO_2 to prevent particle aggregation [22].

However, FSCs can inhibit dissolution, thereby causing novel synthesis mechanisms, such as gas or supercritical anti-solvent (SAS) [23]. These procedures operate at high pressure and involve mixtures between an organic solution and a supercritical fluid [24]. As a consequence of high-pressure dissolution, the solution expands, and the solubility of the solute exhibits decrease rapidly. Finally, the mixture becomes supersaturated, and nanoparticles precipitate [25]. Similarly, particles of saturated gas solutions (PGSS) consist of dissolving a supercritical fluid in a solvent followed by rapid depressurization of this mixture through a nozzle that causes the formation of solid particles or liquid drops [26–29].

FSCs are also utilized in hydrothermal or solvothermal synthesis, participating through chemical reactions, such as hydrolysis and dehydration of solutions of metal salts in subcritical or supercritical water [30–32] or another solvent [33–35]. A novel recently synthesis technique is FSC microemulsion, in which water droplets in sc-CO_2 function as a micro-reactor [36–39].

Therefore, it is important to highlight one of the main advantages offered by technology with supercritical fluids. Through a simple change in operating conditions, it is possible to control the size and shape of the synthesized active phases. In this way it has been possible to produce a wide variety of materials, such as nanoparticles, powder coatings, films, core-shell particles, and nanocomposites.

Furthermore, it is complemented by certain organometallic complexes for the manufacture of different metals and metal oxides [40]. In the 1990s, several pioneering studies on the subject were reported [41–44] which subsequently led to the advancement and development of new processes, which can be summarized in the following fields:

- The use of organo-transition metal complexes as homogeneous catalysts for reactions in SCF and biphasic systems [45–48]
- Extraction of heavy metals from various matrices with SCF [49–51]
- Synthesis of inorganic powders with controlled size distribution by decomposition of organometallic complexes in SCF [52–56]
- Impregnation of polymers with different metal complexes and the subsequent transformation of the metal precursor within such matrices [57–62]
- Incorporation of metallic nanoparticles in porous inorganic and carbonaceous substrates by deposition of SCF [63–65]

This chapter focuses on the last point, in order to contribute to developing the incorporation of metallic active phases in porous substrates by dissolving organometallic complexes in sc-CO_2.

2.2 Supercritical fluid deposition

In recent years, the interest in nanostructured materials has grown enormously [66–69] due to the advantages they present in the area of heterogeneous catalysis. These are associated with its small size and, therefore, with its large specific surface.

In this way, several research studies have focused on improving the synthesis procedures of new catalytic materials, depositing metallic nanoparticles in porous materials (micro-, meso-, or macroporous), with high surface areas and uniform pore structures [70–74].

There are numerous techniques for incorporating metals on substrates, as well as for preparing supported metal films and/or nanoparticles, such as impregnation, coprecipitation, sol–gel, chemical vapor deposition, or microemulsions using organic stabilizing agents. All exhibit different advantages and disadvantages, but, in general, the main drawback of these synthesis methods is the control of the dimensions and dispersion of the particles, as well as the metallic content in the matrix of the support.

Reproducibility, size control, and stabilization under severe operating conditions are the most important challenges in the area of materials synthesis. In view of this, the use of supercritical fluids represents a promising synthesis method for the deposition of metallic nanoparticles on the solid surface and/or inside of porous substrates [75, 76]. Deposition with supercritical fluids (SCFD) is a very simple and ecological procedure [77, 78].

Reproducibility, size control, and stabilization under severe operating conditions are the most important challenges in the area of materials synthesis. In view of this, the use of supercritical fluids represents a promising synthesis method for the deposition of metallic nanoparticles on the surface of solids and/or within porous substrates.

Among SCFs, sc-CO_2 is the most widely used because it is abundant, inexpensive, nonflammable, nontoxic, environmentally benign, and easily accessible and leaves no residues that cannot be treated. Furthermore, it has a critical temperature (Tc) of 31°C and a critical pressure (Pc) of 71 bars.

In addition to the environmental benefits, sc-CO_2 has a high permeability index in almost all polymers, which makes it possible to incorporate metal precursors in various substrates. Furthermore, the dispersion of the particles, the diffusion rates in the substrate, and the separation of the SCF and the polymer chain of the metal of interest can be controlled through changes in temperature and pressure. These good sc-CO_2 properties have also been used in many applications such as separations (extraction and purification), chemical reactions, chromatography, drying of materials, and the synthesis of nanostructured materials [79, 80].

The SCFD technique involves three main stages (**Figure 4**) [81]:

i. Dissolution of the metallic precursor

ii. Adsorption and absorption of the metallic precursor in the support

iii. Reduction of the precursor to its metallic form

Thus, the first stage consists of dissolving an organometallic precursor using sc-CO_2 under optimal conditions to promote solubility in the medium: high pressure and moderate temperature. In the second phase, the resulting solution is exposed to the porous substrate with its subsequent adsorption on the support matrix, which represents the impregnation itself. Then the metal complex can be converted to its metal form as shown in **Figure 4**. Finally, the system decompresses slowly and

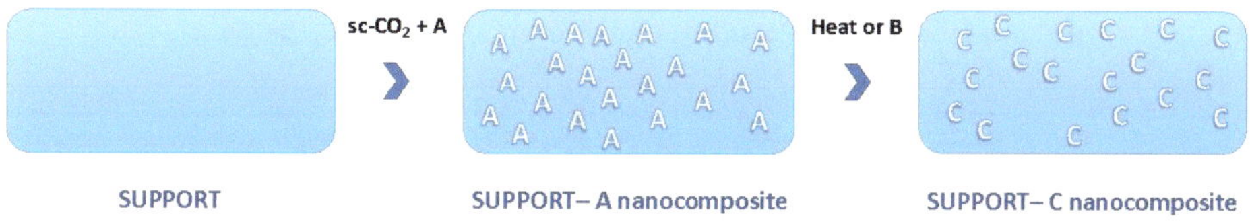

Figure 4.
Schematic diagram for supercritical fluid deposition method. (a) Organometallic precursor, (b) reducing agent, and (c) metal nanoparticles.

gradually until reaching atmospheric pressure. In this way, solvent residues and pore collapse are avoided, the very common difficulties in techniques for preparing catalysts.

The conversion of the adsorbed metal complex into metal nanoparticles can be carried out by different procedures. The conventional method is heating in an inert or hydrogen atmosphere, which can even be done before depressurization in the supercritical fluid phase.

Supercritical fluid reactive deposition (SCFRD) is a term used to refer to the case where metal reduction is performed in the synthesis process [82]. This technique is based on promoting the precipitation of metallic nanoparticles and their consequent deposition and growth on the pore while the organic part of the precursor remains dissolved in the sc-CO_2 and is released together with the gas [66, 83]. In this context, there are two possible ways: (i) injection of a reducing agent, such as H_2 or an alcohol, or (ii) a heat treatment to cause decomposition of the precursor. Consequently, this single-stage process for the synthesis of catalysts makes it possible to reduce operating times compared to other conventional techniques.

2.2.1 Dissolution of the metallic precursor in sc-CO_2

Dissolving the metal precursor in the supercritical phase is the first step in the deposition method. The metallic precursors can be organometallic or inorganic compounds, which will depend on the FSC used in the deposition. In particular, organometallic compounds have significant miscibility in SCFs such as sc-CO_2 and have been extensively used in SCFD techniques. In the case of inorganic precursors such as salts, they only show appreciable solubility in the case of critical and supercritical water [83].

The solubility and behavior of the precursor-SCF phase are very important parameters, since the adsorption of the precursor on the support depends on the concentration in the fluid phase. Some studies on the solubility of organometallic complexes in sc-CO_2 have been reported in the literature. From the point of view of solubility in sc-CO_2, the organometallic complexes can be diketones, dithiocarbamates, macrocycles, organophosphate reagents, hydroxamic acid, and other organic complexes. The organometallic precursors commonly used in the preparation of supported nanocomposites using SCFD are seen in **Table 3**.

2.2.2 Adsorption of the metallic precursor in the support

One of the most important aspects of the technique for preparing supported metals so that the particles are dispersed within the matrix is the absorption of the organometallic precursor in the substrate in the presence of SCF.

Similarly, in the event that the particles must be located on the surface of the support, the adsorption of the organometallic precursor on the surface in the presence of SCF plays an important role. Therefore, the study of the kinetics and

	β-Ketones	β-Fluored ketones	Cyclic ligands
Co			$Co(cp)_2$
Ni			$Ni(cp)_2$
Cu	$Cu(acac)_2$	$Cu(hfac)_2$	
	$Cu(tmhd)_2$	Cu(hfac)	
	$Cu(tmod)_2$		
Ru	$Ru(acac)_3$		$Ru(cod)(tmhd)_2$
Rh	$Rh(acac)_2$		Rh(cod)(acac)
Pd	$Pd(acac)_2$	$Pd(hfac)_2$	PdMe(cp)
Ag		Ag(cod)(hfac); Ag(hfac)(hmte) Ag(hfac)(tegme)	
Pt	$Pt(acac)_2$		$Pt(cod)me_2$
Ir			Ir(cp)(cod)me
Au	$Au(acac)me_2$		

acac—acetylacetonate; cod—1,5-cyclooctadiene; cp—cyclopentadienyl; hfac—hexafluoroacetylacetonate; hmte—hexamethyltriethylene; me—methyl; tegme—tetraethylene glycol dimethyl ether; tmhd—tetramethylheptadionate; tmod—trimethyloctanedione.

Table 3.
Summary of the organometallic precursors used in SCFD [83].

thermodynamics of the absorption and desorption processes is beneficial for the development of this technology.

The resistance to mass transfer on the outer surface of the particle and/or diffusion within the particle governs the kinetics of adsorption or absorption. In the first case, the resistance is quantified by the external mass transfer coefficient of the precursor, while, in the second case, it is a combination of the coefficient of diffusion of the fluid in the pores and of the diffusion of the solute adsorbed on the surface of pores. In absorption, the diffusion of the precursor into the SCF has to be considered when the matrix is full.

3. The importance of catalytic support

Like with metals, choosing a suitable support is a key aspect of catalyst design. The main role of a catalytic support is to fix and disperse the active centers, typically small metal particles or metal oxides, that participate in the reaction of interest. To maximize their efficiency, it is crucial to facilitate close contact between the reactive molecules with the nanoparticle-based surface atoms.

From these postulates, great efforts are made by several research groups to synthesize and optimize chemically stable support synthesis routes, with high and optimal surface areas to disperse active metallic phases.

Due to the high cost of certain metal catalysts such as noble metals, on an industrial scale, small particles with a good distribution and high surface/volume ratio are required.

Although the effective dispersion does not depend on the metal-support interaction, it is known that the morphology and physicochemical characteristics of the supports produce significant effects and can influence the catalytic activities. Carbonaceous materials, alumina and silica, are the commonly used catalytic supports.

3.1 Mesoporous supports

The main difference of mesoporous materials with zeolites is that the walls of their pores are not crystalline because they are made of amorphous silica, which does not have order at the atomic level. Thus, the order of these materials is related to the arrangement of the pores. Mesoporous materials have been studied as possible substitutes for zeolites, although their surface is also polar and they do not have the acidity of zeolites since they do not contain aluminum cations [84]. In addition, the size of their pores is greater than 2 nm, which means that there is no selectivity of form or present intra-particle diffusion problems [85].

Among the most studied mesoporous solids, a breakthrough originated in 1992 when the Mobil Research and Development Corporation published the synthesis of a series of mesoporous silicates that they designated as the M41S family. The main porous solids in this family are named MCM-41, MCM-48, and MCM-50 [Mobil Composition of Matter (MCM)] [86, 87]. The solid MCM-50 has an unstable lamellar structure, while the MCM-48 has a porous cubic metastable structure, which consists of two independent tri-directional channel systems (**Figure 5**).

The structure of MCM-41 silica is made up of a well-defined hexagonal network of unidirectional cylindrical pores, which provides the material with a high specific surface area (~1000 m^2 g^{-1}) and a high pore volume [88, 89].

In general terms, the synthesis of these solids is carried out through the interaction between an inorganic phase and another micellar phase of an organic nature. The precursor agents of the structure or surfactants (templates) are characterized by being molecules with an amphiphilic character or also called "amphipathic molecules." In other words, they have a hydrophilic end (soluble in water) and a hydrophobic end (rejects water), which is usually a hydrocarbon chain. Like any other surfactant, they form micelles when in aqueous solution under certain conditions because they have a hydrophobic and a hydrophilic chain that cannot be separated. The surfactant micelles serve as templates or molds and create the internal porosity of the solid.

The micellar arrangement, responsible for the structuring of the final material, was attributed to the presence of surfactant molecules in an aqueous medium, under certain conditions of temperature, pH, and concentration, forming ordered structures known as liquid crystals [90].

The mechanism of formation of mesoporous materials is generally carried out through the sol-gel process, which has a number of advantages over others such as low temperature and homogeneity at the molecular level.

This method usually involves the hydrolysis and condensation of the precursors, which can be metal alkoxides or inorganic or organic salts. Furthermore, organic or

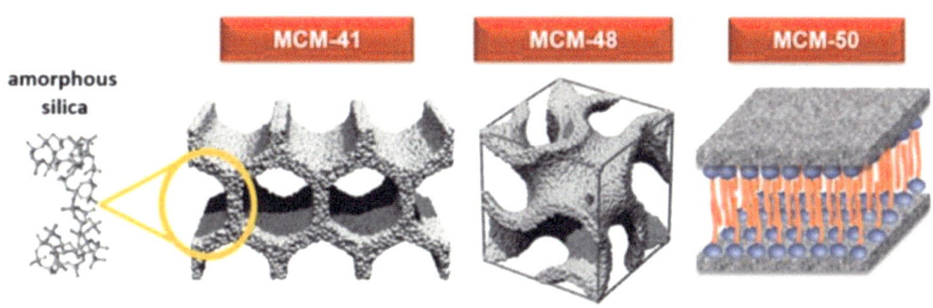

Figure 5.
Members of the M41S family. Adapted from [161].

aqueous solvents are used to dissolve the precursors. Occasionally, other substances are used to promote hydrolysis and condensation reactions [91]. Normally, the alkoxide groups are chosen in a sol-gel synthesis process as the silicon source. The reactivity of the chosen precursors largely depends on the degree of charge transfer and the ability to increase their coordination number.

Hydrolysis and condensation reactions are multistep processes, occurring sequentially, in parallel, and may be reversible. Condensation results in the formation of metal oxides or hydroxides, often with attached or attached organic groups. These organic groups may be due to incomplete hydrolysis or introduced as non-hydrolysable organic ligands [92].

The dimension and size of the pore system are dictated by the dimension and size of the micelles in solution. In the case of the hexagonal MCM-41 material, the pore diameter is between 20 and 40 Å when the length of the alkyl chain in the surfactant varies between C8 and C16. Finally, by calcination, excess surfactant is removed.

The siliceous walls of these materials do not have any order and are full of structural defects resulting from the hydrolysis of the silica source and its subsequent condensation. This condensation is not complete, so silicon atoms remain attached to OH– groups on the walls, called silanol groups.

3.1.1 Synthesis of mesoporous support

In general terms, the synthesis of mesoporous solids is carried out through the interaction between an inorganic phase and another micellar phase of an organic nature. The precursor agents of the structure (surfactants or templates) have a hydrophilic end (soluble in water) and a hydrophobic chain (rejects water), which is usually a hydrocarbon chain. Generally, these two parts tend to separate if added to a mixture with other substances, which cannot be fulfilled because both ends are joined by a chemical bond.

3.1.1.1 MCM-48

This support was prepared using a conventional hydrothermal method. For that, n-hexadecyltrimethylammonium bromide (2 g) was used as template [$CH_3(CH_2)_{15}N(Br)(CH_3)_3 \geq 98\%$, Sigma-Aldrich], ammonium hydroxide to condition the basic medium (13 cm^3, 20% as NH_3), and absolute ethanol as solvent (18 cm^3). The subsequent solution was stirred for 15 min, and the tetraethylorthosilicate (TEOS) was added such as Si source. It is necessary that TEOS were added dropwise to avoid the quick condensation. Then, the solution was further stirred for 18 hours at 30°C. Afterward, the white precipitate was collected by filtration, washed with distilled water, and dried at 60°C overnight.

Finally, the template was removed by calcination with a heating rate of 2°C min^{-1} at 550°C and maintained at 550°C for 8 hours in airflow.

3.1.1.2 MCM-41

The mesoporous substrates MCM-41 and AlMCM-41 were synthesized in a spherical way. For this, the commercial reagents that were used were the following:

- Surfactant: n-hexadecyltrimethylammonium bromide (C_{16}TMABr)
- Silicon source: tetraethyl orthosilicate (TEOS)
- Aluminum source: sodium aluminate ($NaAlO_4$)

- Ethanol (EtOH)
- Ammonium hydroxide solution (NH_4OH) 29% v/v

Thus, 1 g of C_{16}TMABr was dissolved in 19.2 ml of water at room temperature and continuous stirring. Then 24 g of absolute ethanol and 5.9 g of NH_4OH solution were added. Subsequently, 1.88 g of TEOS was added dropwise, to avoid rapid condensation of the Si source.

Ethanol is added to the preparation because it acts as a solvent, while water is a reagent for the hydrolysis reaction of TEOS. Ammonium hydroxide is used to achieve an alkaline medium and thus obtain discrete support particles. The mixture was kept under continuous stirring for 18 hours at 25°C. The resulting solid has the following molar composition:

$$\text{TEOS:0.3 } C_{16}\text{TMABr:11 } NH_3\text{:144 } H_2O\text{:58 EtOH}$$

The AlMCM-41 sample was prepared similarly to the MCM-41 solid. The difference is the addition of 0.04 g of sodium aluminate before adding TEOS, with the ratio Si/Al = 20.

The final solid was obtained by vacuum filtration, with subsequent washes until neutral pH was reached. Then, the samples were dried for 24 h at 80°C, and, finally, they were calcined in airflow at 5°C min^{-1} up to 550°C to achieve the total elimination of the surfactant.

4. Metals in catalysis and their applications

The area of heterogeneous catalysis is constantly developing, establishing innovative synthesis methods and modern metal deposition techniques, creating original supports, or trying to discover the necessary physicochemical properties of the catalyst for each reaction.

In this context, noble metal-based catalyst systems exhibit high activities for most reactions and are commonly used in numerous industrial applications. The main problem with the use of noble metals as the active phase is that, despite acting as an inert reaction, over time they deactivate due to poisoning with contaminating substances, coke deposition, or leaching of the active phases.

From this perspective, the development and extensive use of catalysts based on non-noble metals, such as Ru, Co, and Ni, are considered to be preferred from an industrial point of view, mainly for economic reasons.

4.1 Cobalt-based catalysts

In heterogeneous catalysis, cobalt represents one of the most studied active phases for decades. In particular, 2700 tn Co(0) are used annually as a homogeneous or heterogeneous catalyst in petrochemicals or plastic industries. Cobalt is used in the production of ultra-resistant resins or plastics for different purposes.

In addition, Co is used for the production of alcohols and aldehydes and consequently obtaining detergents. The active species can be metallic, cobalt oxide, hydroxide, or ionic.

On the other hand, there are numerous articles that report the behavior of cobalt-based catalysts applied in a wide variety of reactions of industrial significance. Those involving the breaking of C–H and C–C bonds at low temperatures are important. In this sense, cobalt catalysts supported on zirconia, ceria, and alumina have reported high catalytic activity, similar to that obtained with the noble metal [93–96]. These solids have also been used satisfactorily in the dry-steam reforming process [97] in order to obtain synthesis gas enriched with hydrogen.

Llorca [98] demonstrated that the dispersion and reducibility of the cobalt species together with the cobalt-support interaction have important effects on the catalytic performance, being the key parameters in the stability and selectivity of the process. In order to improve efficiency, the synergistic effect of cobalt with other metals such as Ni [99–101], Ru [102, 103], and Fe or Cu [104] is also investigated.

Furthermore, there is a large group of research that focus on developing new materials for the Fischer-Tropsch synthesis (FTS) [105–107]. In order to efficiently obtain hydrocarbon mixtures from syngas (CO_2 and H_2) derived from coal, natural gas, or biomass resources, cobalt-based catalysts are widely used in this type of process due to their selectivity towards long-chain paraffin and strong resistance to deactivation and because they are economically viable. The nature and structure of the catalytic support, as well as the dispersion of Co(0) metal sites and the reducibility of cobalt species, directly influence the FTS performance. As a consequence, several synthesis strategies have been studied in recent years.

In this context, in order to achieve a high dispersion of surface active cobalt species, in some cases metal-support interaction is to stabilize the catalyst. Therefore, different supports have been studied such as titanium or magnesium oxides, zeolites, mesoporous silica (MCM-41 and SBA-15) [106, 108, 109], and carbon nanostructures (CNFs, MWCNTs, and CSc) [105, 110].

However, a strong interaction with the support can also decrease reducibility by creating nonreducible Co species such as $CoAl_2O_4$ and $CoTiO_3$. For this reason, some noble metal promoters, such as Pt, Re, or Ru [111, 112], are used to optimize the reduction of cobalt oxides and to inhibit coke deposition during FTS.

Although these are the best-known application processes, there are other examples where promising developments are taking place.

Unsupported cobalt catalysts showed good activity for ammonia synthesis [113, 114], improving thermal stability when cerium or barium were incorporated as promoters.

On the other hand, there is great scientific and environmental interest in reducing or eliminating NO_x emissions from combustion gases by means of selective catalytic reduction (SCR-NO_x). Cobalt catalysts supported on zeolites (MFI or FER) showed an increase in catalytic activity directly related to metal charge [115, 116].

In this sense, great efforts are made to develop new catalysts based on Co and applied in the combustion of volatile organic compounds (VOCs) generated by industry, transports, and residential/service sector [117–119]. To this point, high catalytic yields have been reported for the VOC oxidation at low temperatures given by cobalt oxides such as Co_2O_3 and Co_3O_4. This carries considerable environmental and economic benefits compared to traditional thermal oxidation.

4.1.1 Deposition of cobalt using sc-CO_2

Hunde and Watkins [120] were pioneers in the synthesis of cobalt and nickel on native silicon oxide using supercritical fluids. They also reported on the formation of Si-supported TaN and TiN films, without the need for a catalytic layer, by direct reduction of H_2 present in the cobaltocene ($CoCp_2$). For this, it was operated at a reduction temperature of cobaltocene in the range between 285 and 320°C and a pressure between 22.0 and 26.0 MPa. No significant deposition below 280°C was observed.

4.2 Nickel-based catalysts

Nickel-based catalysts are widely used due to their high activity at room temperature, leading to participation in the competitiveness of several industries.

The most important applications of this active phase include artificial fibers, vehicles parts, and nylon production. Adipic acid and caprolactam are obtained from the process of transforming benzene into cyclohexane using a nickel-based catalyst [121].

Also, a large part of the production of oils and fats derived from natural sources and used as food products (palm and vegetable oils) depends on the utilization of nickel catalysts. Fertilizer production involves technological processes with nickel catalysts for the ammonia production.

However, its main application is in oil refining. Raney Ni catalyst is widely used in hydrodenitrogenation to reduce NO_X and hydrodesulfurization to decrease SO_X concentration so that it is possible to obtain concentrations of these pollutants with levels lower than those legally required.

In addition to these applications, nickel catalysts supported on aluminum and porous silica represent the preferred option for cracking and hydroprocessing due to their ability to adsorb large amounts of hydrogen, which increases the efficiency of the reactions. It is important to note that nickel is less expensive relative to other metals like platinum.

Another field of application is in the production of synthesis gas as an alternative to the use of fossil fuels. One of those production methods is the reforming with liquefied natural gas [122, 123], ethanol [124], or an aqueous fraction of bio-oil [125, 126]. Nickel catalysts showed a good yield towards hydrogen production at a considerably lower cost than other noble metals in both fixed-bed and fluidized-bed reactors. However, these catalysts have the disadvantage of being deactivated significantly due to carbon deposition and sintering due to extremely severe operating conditions, leading to loss of active surface. For these reasons, there are continuous studies seeking to solve this problem facing it from different perspectives: the use of promoters (Zn, Mg, Zr, others [124, 126]), oxide supports (SiO_2 or Al_2O_3 [123, 127]), and different recuperation treatments [128].

Nevertheless, the most attractive process for syngas production is the CO_2 reforming of CH_4 or dry reforming (DRM) and all the side reactions also catalyzed by nickel.

These types of catalysts are replacing noble metals, such as Ru, Rh, and Pt, with DRM due to their availability and profitable viability. But, once again, its use is limited due to its high tendency to coke deposition and consequent deactivation. Side reactions, including CH_4 cracking [129], lead to the generation of carbonaceous deposits [130]. Furthermore, the endothermic nature of the reaction indicates that it is necessary to operate at high temperatures, tending to metal sintering. In this regard, the main parameters to be taken into account to reduce coke formation and improve catalytic performance have been studied and published, such as the nature of the support [131–133], the catalyst preparation method [134, 135], and the presence of modifiers [136, 137].

At the same time, among the growing concern over the depletion of fossil fuel reserves and environmental pollution, the generation of hydrogen from renewable resources such as biomass is emerging as an attractive alternative to energy demands. The use of nickel catalysts in the biomass gasification process cannot only increase H_2 yield but also decrease reactor size as a result of optimizing the reaction rate. Due to the same difficulties mentioned above, various catalytic supports, such as alumina, silica, and magnesium, were developed with metallic additives to improve thermal stability like Rh, Ru, Co, Fe, and Cu [138, 139].

Another instance is CO_2 methanation, developed by Sabatier and Senderens [140], which was industrialized in 1970 thanks to the use of catalysts, mainly Rh, Ru, or Pd. Recently, several supported nickel-based catalysts have been tested with high selectivity towards methane and a relatively low price [141, 142].

In addition, there are several articles and patents [143, 144] on the use of nickel as a catalyst for the hydrogenation of organic compounds, due to its ability to mainly hydrogenate C=C double bonds, carbonyl groups (aldehyde and ketone), nitro, nitrile, and oxime.

4.2.1 Deposition of nickel using sc-CO_2

Up to now, there are few articles related to the deposition of Ni nanoparticles by SCFD on micro- and mesoporous supports.

Peng et al. [145] have shown that $NiCp_2$ hydrogenolysis can take place by means of a low temperature autocatalyzed process (<70°C) in sc-CO_2 in both carbon nanotubes and flat surfaces.

In this sense, Bozbag et al. [146] reported the synthesis of Ni/AC, using activated carbon and $Ni(acac)_2$ as a precursor. The sc-CO_2 was used at 30 MPa and 60°C, followed by a heat treatment using H_2 at atmospheric pressure, achieving metallic loads of up to 6.5 wt.% with an average nanoparticle size around 10 nm.

Similarly, Taylor et al. [147] prepared Ni catalysts supported on carbon nanotubes and aluminosilicates with sc-MeOH and $Ni(acac)_2$, obtaining higher metal loads (60–70 wt.%). These were used in the production of hydrogen through the gasification of biomass in supercritical water.

Finally, the formation of a $NiCp_2$ film on Si, TaN, and TiN by means of chemical fluid at temperatures between 175 and 200°C and pressures from 19 to 23 MPa was reported [120]. The films were found to be essentially free of ligand-derived contamination, using a high-pressure cold-wall reactor and, therefore, restricting deposition only to heated substrates.

4.3 Ruthenium-based catalysts

Ruthenium is a transition metal that belongs to group VIII and has properties similar to those of the platinum. Ru-based catalysts can have a high potential in oxidative catalytic reactions such as asymmetric epoxidation of alkenes, dihydroxylation of olefins, or oxidative dehydrogenation of alcohols.

Ruthenium catalysts, mainly known industrially as Grubbs, are widely used in olefin metathesis reactions. They are widely applicable and requested due to their high tolerance to functional groups and stability [148, 149].

Similarly, in fine chemistry and pharmacy, Ru-based solids have been developed that facilitate hydrogen transfer. This alternative, which consists of adding hydrogen to a molecule from a source other than gaseous H_2 in the presence of a catalyst, is increasingly being used in the area of organic industry. On the other hand, ruthenium complexes are used in the reduction of ketones, aldehydes and imines to produce alcohols and amines, respectively [150, 151].

Many studies are trying to replace iron and other transition metals in reactions that require high temperature and pressure to achieve sufficient productivity. This can be seen in the synthesis of ammonia, for example, for which various ruthenium catalysts supported on $BaTiO_2$ [152] and carbon [153] have been tested.

In the energy field, a great deal of effort is devoted to finding materials that guarantee a safe and economical way to store hydrogen. Ruthenium-based catalysts are earning a significant place as a result of high yields. Akbayrak et al. [154, 155] have reported that ruthenium supported on silica or nanotitania is one of the most active catalysts for the hydrogen generation from ammonia borane under mild conditions.

However, in general, most of the work focuses on hydrogenation reactions [156–158]. For example, in the industrial hydrogenation of benzene to obtain cyclohexane, nickel-based catalysts are used at temperature above 200°C and below 50 bars. These operating conditions lead to poor performance and promote undesired side reactions such as isomerization and hydrocracking, thus decreasing selectivity towards cyclohexane. In view of these arguments, numerous researchers are trying to replace this type of catalysts by others based on ruthenium alloys [156, 157].

Furthermore, as mentioned above, Raney nickel catalysts are used in the hydrogenation of glucose. However, new ruthenium-based catalysts are being developed due to their resistance to chemical attack [158, 159].

Increasingly, new types of ruthenium-based catalysts catalyze multiple hydrogenation reactions, even under severe operating conditions, such as transforming CO_2 into formic acid below 20–40 MPa and 80–150°C [160].

4.3.1 Deposition of ruthenium using sc-CO_2

There are some articles that study ruthenium deposition on different micro- and mesoporous supports using supercritical fluids. In this context, Yen et al. [161] showed the synthesis of Ru-MCM-41 solids using Ru(cod)(thmd)$_2$ as a precursor at 150°C, 10 MPa of CO_2, and 10 MPa of H_2. In this way, metallic nanoparticles are effectively distributed and dispersed with an average diameter of 3.4 nm. Likewise, Kim et al. [162] reported methods for preparing ruthenium nanodots on Si and HfO_2 films. In this case, $Ru_3(CO)_{12}$ was used as a precursor and dissolved in sc-CO_2 for 1 h, at 90°C and 23 MPa.

On the other hand, activated carbon is one of the most used catalytic supports in recent times. Due to this, there are numerous works that analyze the deposition of ruthenium on activated carbon [163–165] with sc-CO_2 and cosolvents such as methanol and ethanol. Operation variables such as temperature, pressure, and amount of cosolvent are optimized, being 45°C and 10 MPa the optimal conditions to obtain 2 wt.% of ruthenium.

5. Catalyst synthesis by SCFD

In this section, the supercritical fluid deposition will be developed for the synthesis of cobalt-, nickel-, and ruthenium-based catalysts on mesoporous supports using batch and continuous systems and detailing different configurations of the high-pressure reactor.

5.1 Batch device

The batch deposition experiments were carried out in a stainless-steel high-pressure vessel with an internal volume of 100 ml. The experimental setup operates up to 30 MPa pressure and 400°C maximum allowable working conditions (**Figure 6**).

A Milton Roy Dosapro metering pump (flow rate up to 6.2 L h^{-1} and a maximum pressure of 33 MPa) was used to supply the CO_2 to the reactor. The fluid is pre-cooled in a bath with an ethylene glycol/water (50/50) mixture cooled with an immersion cooler to ensure the liquid state of CO_2 for pumping.

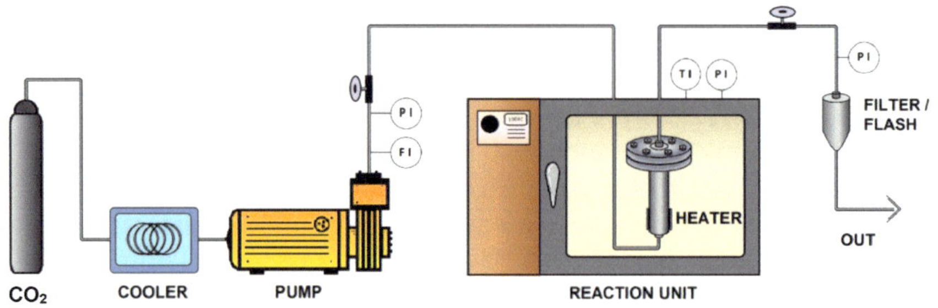

Figure 6.
Experimental device for the catalyst synthesis by batch SCFD.

Mass flow meter and a pressure gauge are installed in the pump outlet. The support and organometallic precursor are placed separated by a metal mesh to facilitate the circulation of sc-CO_2 and avoid its direct contact (**Figure** 7).

The reactor is equipped with two wall-mounted electrical resistors located to promote convective flow of sc-CO_2 and a K thermocouple.

During the first step, a precursor is dissolved by sc-CO_2 under optimal operation conditions to favor its solubility. Meanwhile, precursor adsorption onto the support from the medium takes place during the desired adsorption time. After this adsorption step, temperature is increased to 200°C to break down the organometallic compound, with its subsequent precipitation of the metallic nanoparticles during the desired decomposition time. Afterwards, the system is isochorically cooled down to subcritical conditions, and CO_2 is released from the reactor over a period approximately 30 min, which implies a slowly depressurization to atmospheric pressure to avoid drag metallic nanoparticles which are adsorbed over the support surface.

5.1.1 Mass transfer limitations

Due to the configuration of the experimental device, the synthesis process is carried out into the reactor without external stirring. To facilitate as far as possible the homogeneity of the solution, the electrical resistances are located at the bottom of the vessel, as it is shown in **Figure** 7, to favor the creation of convective currents.

Inside the high-pressure reactor, the support and precursor are located inside two glass vials separated by means of a wire mesh. The precursor is located in the lower part, and the support occupies the upper vial. This kind of configuration has been called "column" type.

Since the support is disposed as a fixed bed, it was decided to test whether all along the vial the same deposition was obtained or conversely there was some mass transfer limitation. For this reason, the metal content from the same sample was analyzed at three different heights. The results showed that there was a certain divergence, with a concentration gradient of up to 20% if the initial amount of support was too high, so a new configuration in the form of placing the reactants was contemplated in order to avoid the observed mass transfer limitations.

To avoid bed length which was too long, reagents were separated into smaller quantities in several vials and placed alternately, maintaining the total mass of both the precursor and support constant. This new configuration is called "sandwich," and a scheme of it is presented in **Figure 8**. The results are compared in **Table 4**.

Surprisingly, the improvement allows obtaining nanocomposites with really higher metallic loadings to 16 wt.%, showing that the "sandwich" configuration improves mass transfer in the system.

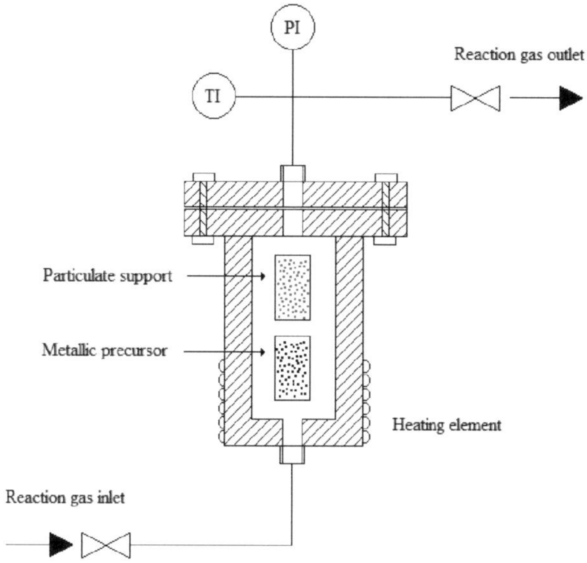

Figure 7.
Distribution of catalytic support and organometallic precursor within deposition chamber.

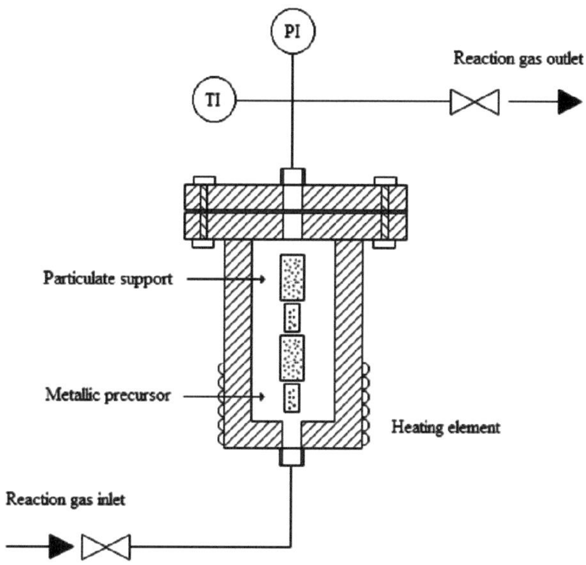

Figure 8.
Distribution of catalytic support and organometallic precursor in a "sandwich" configuration.

5.2 Continuous setup

The SCFD in continuous will be developed for synthesis of ruthenium-based catalysts. The continuous system is represented in **Figure 9**. CO_2 is pumped with a Jasco model PU-2080 Plus HPLC pump, which operates with a flow range between 1 μL to 10 ml min^{-1} and a maximum pressure of 22 MPa. The pump extraction line is cooled to guarantee the liquid state of the CO_2. So as to maintain a constant flow throughout the experiments, a mass flow meter controller with Bronkhorst Coriolis sensor calibrated to CO_2 combined with a controller model mini Cori-Flow (10–100 g h^{-1}) was established. In addition, a back-pressure valve is placed downstream of the reactor to control and manipulate the desired pressure. Finally, a Swagelok

Batches	wt.% Co	
	Column	Sandwich
1	3.2	5.8
2	4.4	13.7
3	6.0	16.6

Table 4.
Cobalt loading after each successive batch in column or sandwich configuration.

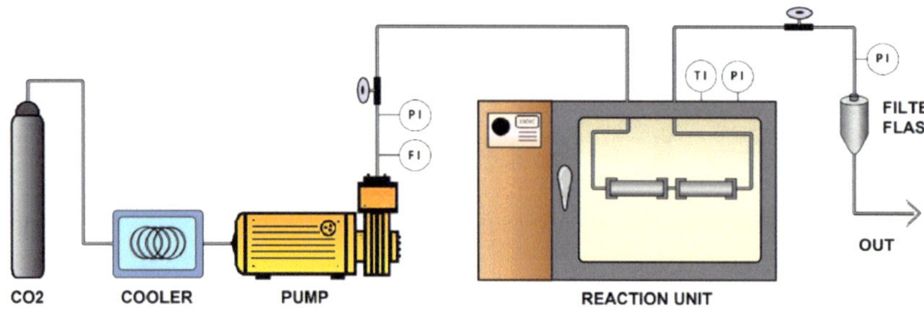

Figure 9.
Experimental device for the synthesis by SFRD in continuous mode.

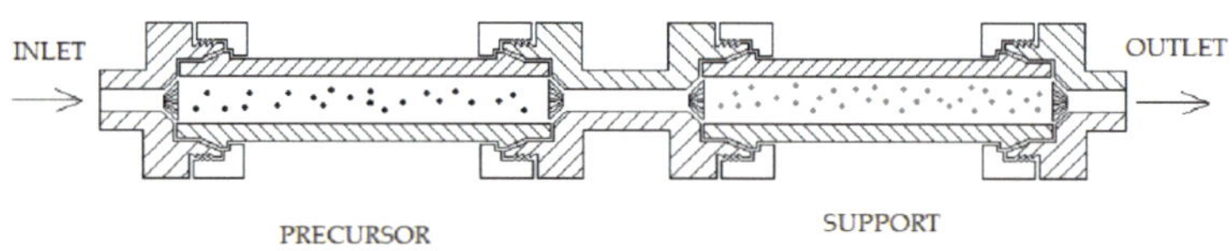

Figure 10.
Internal configuration of the high-pressure reactor in the continuous SCFD.

SS-4R3A relief valve is located to protect the system from overpressures, with a nominal pressure of up to 41 MPa at room temperature.

In this new configuration, the high-pressure container used in batch operation has been replaced by two sections located in horizontal position inside the furnace, as shown in **Figures 9** and **10**.

The first piece contains the organometallic precursor mixed with glass spheres to relief its dissolution into the CO_2 flow. Afterwards, the solution goes through the second tube, where the support is located.

The depositions were carried out under controlled conditions of pressure at 11 MPa and temperature at 60°C. The initial flow of CO_2 used (45 g h^{-1}) was sufficient to dissolve the precursor and be saturated before contact with the catalytic support. This premise was based on the observations made in the experiments carried out in the visual cell.

6. Metallic species deposited with sc-CO_2

6.1 Co-MCM-41 and Co-Al-MCM-41 catalysts

6.1.1 Physical, chemical, and textural properties

The adsorption and desorption experiments of N_2 at −196°C together with the measures of small-angle X-ray scattering (SAXS) were used to study the effect of sc-CO_2 and the addition of Co on the hexagonal arrangement of the mesopores and pore size distribution in the prepared samples. **Figure 11** presents the N_2 adsorption/desorption isotherms obtained for the mesoporous support and the cobalt samples. All solids showed type IV adsorption isotherms, typical of mesoporous materials with a strong inflection at relative pressures $P/P_0 > 0.3$, indicating the uniformity of the mesopore size distribution.

Table 5 presents the quantitative results of N_2 adsorption and desorption for the samples with high and low cobalt content, which are compared with those obtained in their respective supports. In addition, the interplanar distance values d_{100} and unit cell parameter a_0 are shown, obtained by SAXS. The interplanar distance "d100," in the direction (1 0 0), was calculated using Bragg's law ($\lambda = 2 \cdot d_{hkl} \cdot \sin\theta$). Also, the unit cell parameter "a_0" was determined, which indicates the distance between the center of two adjacent pores in the hexagonal structure ($a_0 = 2 \cdot d_{100}/\sqrt{3}$) [166, 167]. Both parameters are included in **Table 5**, together with the average pore size of the synthesized samples, obtained from the BJH method.

The addition of aluminum to the MCM-41 support produces an approximate 38% of decrease in BET surface, while the average pore size remains relatively constant between 4.6 and 4.8 nm. On the other hand, beginning from the same amount of precursor, the Co content deposited in the Al-MCM-41 support is slightly higher than in the MCM-41 sample. It is known that the incorporation of Al to pure silica is done to give more acidity to the MCM-41 structure. Therefore, it is probably that there is a greater interaction as a consequence of the aluminum aggregate [92, 168].

Regarding the cobalt incorporation, the specific area decreases 9 and 20%, with the addition of 0.6 and 4.3 wt.% of Co to the MCM-41 support, respectively. On the other hand, the catalysts Co(0.8)-Al-MCM-41 and Co(5)-Al-MCM-41 show a decrease in the area of 7 and 12%, respectively. When the added Co content is high (4.3 and 5 wt.%), there is a slight decrease in the size of the porous cavities, as a consequence of the incorporation of Co inside the mesopores.

Consequently, to determine if cobalt incorporation affects the hexagonal arrangement of mesoporous channels, SAXS technique was used. In the case of amorphous materials such as MCM-41 and AlMCM-41, the regular arrangement or

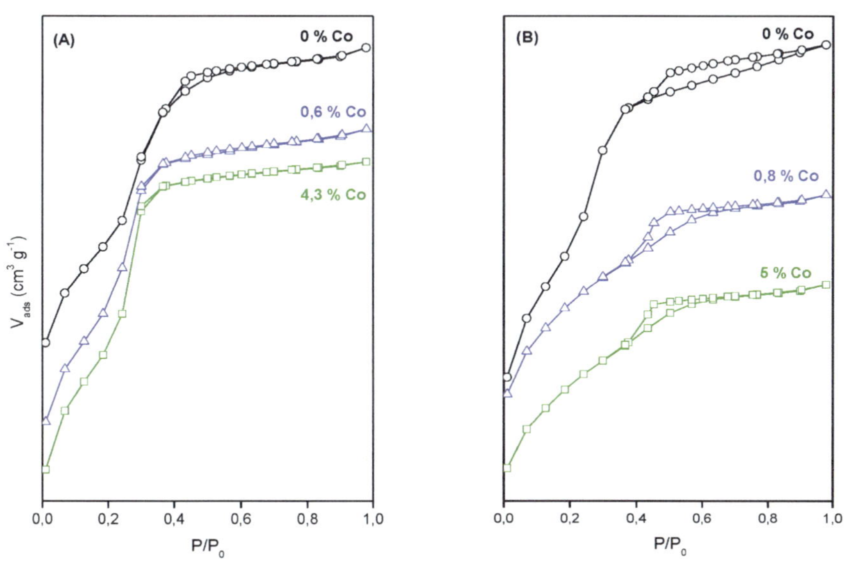

Figure 11.
Adsorption/desorption isotherms of N_2 at −196°C obtained for samples functionalized with Co on supports (A) MCM-41 and (B) Al-MCM-41.

Catalysts	% Co[a]	S_{BET} ($m^2 g^{-1}$)	v_p ($cm^3 g^{-1}$)	t_p[b]	d_{100}[c]	a_0[d]
MCM-41	0	1295	0.194	4.6	3.7	4.3
Co(0.6)	0.63	1183	0.173	4.5	3.7	4.3
Co(4.3)	4.34	1034	0.085	3.9	3.7	4.3
Al-MCM-41	0	807	0.216	4.8	3.7	4.3
Co(0.8)	0.82	752	0.198	4.1	3.4	3.9
Co(5)	5.10	709	0.156	3.9	3.4	3.9

[a]*Determined by ICP (wt.%).*
[b]*Average pore size (nm).*
[c]*Interplanar distance, $d_{100} = \lambda/2 \cdot sen\ \theta$ (nm).*
[d]*Unit cell parameter, $a_0 = 11{,}547 \cdot d_{100}$ (nm).*

Table 5.
Chemical, physical, and textural properties of co-mesoporous catalysts.

ordering of the pores produces reflections that appear as signals at low diffraction angles. The solids of the M41S family have easily identifiable diffractograms providing reflections (h k 0).

The results obtained by SAXS for the Co(x)-MCM-41 and Co(x)-Al-MCM-41 materials are presented in **Figure 12**. The appearance of an intense diffraction peak at 2.4° is characteristic of the (1 0 0) plane of the supports and indicates an ordered porous structure. In addition, there are other weaker diffraction peaks at 4.3 and 5.0° that correspond to the planes (1 1 0) and (2 0 0) and that verify the synthesized mesoporous structure.

As seen in **Figure 12A**, there are no notable changes in the shape and position of the diffraction peaks for the Co(x)-MCM-41 samples. On the other hand, when 5 wt.% of Co is deposited on the Al-MCM-41 substrate (**Figure 12B**), a slight shift of the main peak occurs towards higher diffraction angles ($2\theta = 2.6°$). This suggests that the deposited cobalt charge causes a slight loss of the hexagonal arrangement. Furthermore, the cobalt incorporation in both mesoporous supports produces a decrease in the intensity of the diffraction peaks, which indicates the disorder of the hexagonal arrangement, but not breakage of the pores [87].

6.1.2 Morphology

Figure 13 presents the SEM images obtained after the addition of cobalt on supports MCM-41 and Al-MCM-41 with sc-CO_2 in batch conditions with "sandwich" reactor configuration (see **Table 4**).

It is observed that the procedure using supercritical fluids is a nonaggressive methodology, since the spherical morphology of the catalytic supports is maintained, with an average diameter close to 500 nm. Furthermore, it should be noted that the individual deposited nanoparticles cannot be detected directly using the secondary electron mode. However, when images are obtained through backscattered electron mode (BSE), areas of high electron density are observed, corresponding to cobalt particles. In **Figure 13**, brighter regions are highlighted, indicating areas of high electron density, due to the presence of cobalt. These zones are well dispersed and correspond to cobalt nanoparticles. Using EDX, cobalt concentrations close to 5.9 and 5.1 wt.% were obtained for the Co(4.3)-MCM-41 and Co(5)-Al-MCM-41 catalysts, respectively.

Figure 14 presents the mapping of SEM images from sample Co-MCM-41. Different shades in the gray scale are associated with various emission lines. In this

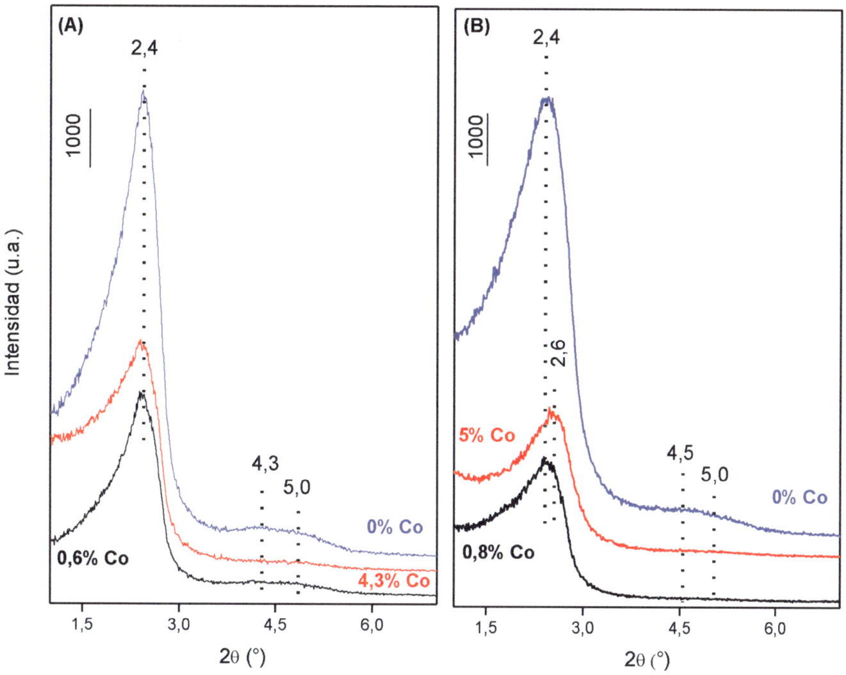

Figure 12.
Results obtained by SAXS for co-functionalized on supports (A) MCM-41 and (B) Al-MCM-41.

way, each tone in the image refers to the unique energy emission of the element of interest. When the cobalt charge was 4.3 wt.%, internal Co species and spherical nanoparticles were seen on the external surface of the MCM-41 substrate. When the concentration of cobaltocene (organometallic precursor) used was greater than the calculated solubility limit (0.336 g L^{-1}), the Co particles obtained had a diameter between 10 and 20 nm. Therefore, during the decomposition phase, the deposition of the nanoparticles occurs on the outer surface and inside the spherical particle of the support.

6.2 Ni-MCM-48 catalysts

6.2.1 Physical, chemical, and textural properties

The content of deposited nickel with sc-CO_2 was close to 3 wt.%, and it was determined by ICP. **Table 6** presents the surface and pore structure parameter of calcined MCM-48 and Ni-deposited catalyst. The BET surface area of the silica support MCM-48 is higher than the MCM-41, and this was 1641 m^2 g^{-1}. The pore volume and diameter were 0.9 cm^3 g^{-1} and 3.6 nm, respectively. After incorporating 3 wt.% Ni with sc-CO_2, the BET area and pore volume decrease close to 40.5%. The loss of textural properties is further accentuated compared to the catalyst with 5 wt.% Co. This can be associated with the fact that the ionic radius of nickel (0.78 Å) is higher than of Co (0.63 Å) and the solubility of nickel precursor (nickelocene) in sc-CO_2 is up to 1.5 times less than that of cobaltocene.

The N_2 adsorption/desorption isotherms (**Figure 15**) of the mesoporous MCM-48 and Ni-MCM-48 show the typical features of a mesoporous silica material, and it can be classified as type IV according to the IUPAC [169]. First, a sharp nitrogen uptake at P/P_0 in the range of 0–0.02 due to a monolayer adsorption on the walls of

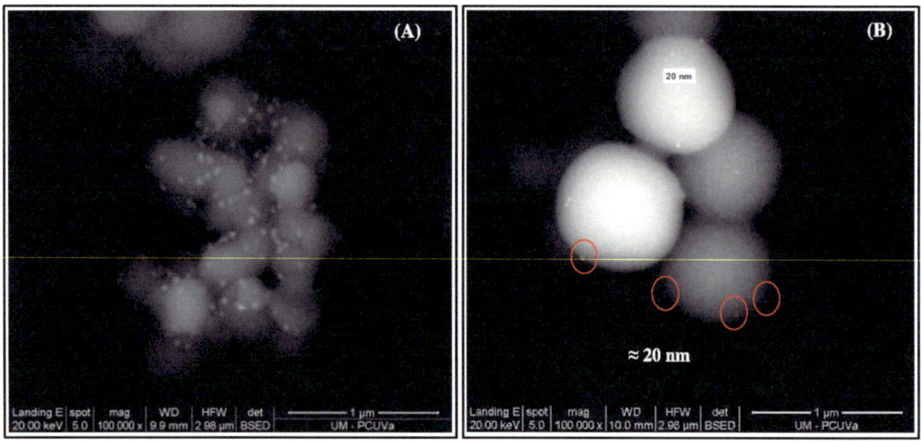

Figure 13.
SEM images obtained by backscattering of (A) Co-Al-MCM-41 and (B) Co-MCM-41.

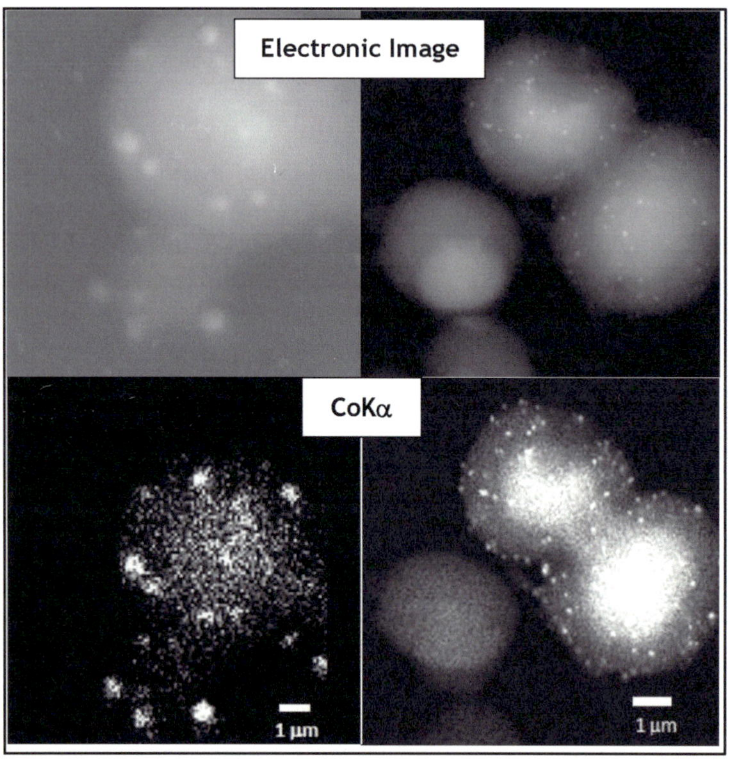

Figure 14.
Mapping of the SEM images obtained for the Co-MCM-41 catalyst.

Catalysts	% Ni[a]	S_{BET} ($m^2 \cdot g^{-1}$)	v_p ($cm^3 \cdot g^{-1}$)	t_p[b]
MCM-48	0	1641	0.90	3.6
Ni-MCM-48	3.0	977	0.45	1.4

[a] *Determined by ICP (wt.%).*
[b] *Average pore size (nm).*

Table 6.
Chemical and textural properties of Ni-MCM-48 catalyst.

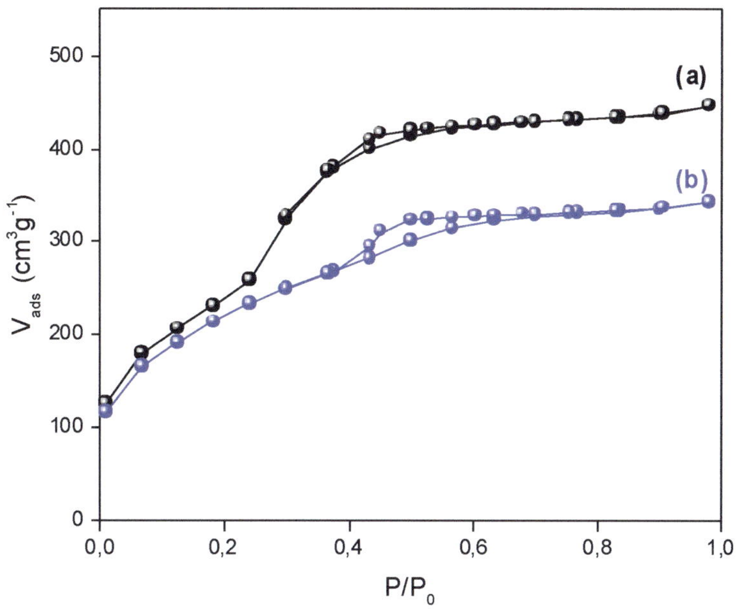

Figure 15.
Adsorption/desorption isotherms of N_2 at −196°C obtained for samples (a) MCM-48 and (b) Ni-MCM-48.

MCM-48 is observed. This step is followed by an abrupt increase in the volume of nitrogen adsorbed at P/P_0 in the range of 0.2–0.3 associated to capillary condensation of N_2 in the channels of MCM-48, suggesting uniformity of the channels and a narrow pore size distribution [170].

6.2.2 Morphology

Table 7 shows the results of nickel content after the impregnation by SCFD in batch conditions with column and sandwich reactor configuration.

The first three experiments were carried out under the same conditions: 14 MPa, 200°C, 1 h, and a reactor configuration in column. The nickel concentration in all of them was very similar, with an average value of 2.7 wt.% of nickel and a standard deviation lower than 0.1 wt.%, indicating the robustness of the SCFD process.

On the other hand, when studying the sandwich-type configuration of the reactor, in which several vials with precursor and support are placed alternately inside the reactor, they do not reveal a significant difference in the metallic charge as if it occurred with cobalt. Although depositions made in sequential charges have achieved increasing concentrations of metal, the rate of increase is considerably less from batch to batch, compared to cobalt. As observed in **Table 7**, the increase in nickel concentration after the second deposition is 52%, while it decreases to 14% after the third batch.

However, it is important to note that this type of configuration allows obtaining a greater amount of catalyst in a single operation.

In this way, it is observed that the SCFD method represents a valuable and versatile technology, with reasonable reproducibility, that can be adapted and applied to other metal-support systems with acceptable efficiencies.

Instead, an electron microscopy (SEM/TEM) was used in order to study the morphology and the dispersion of the nickel nanoparticles incorporated by SCFD in batch and sandwich configuration. In this sense, **Figure 16** shows spherical particles of support with an average diameter of about 200 nm and metallic nanoparticles dispersed on the surface.

Batches	wt.% Ni column	Batches	wt.% Ni sandwich
1	2.8	1	2.5
1	2.8	2	3.7
1	2.6	3	4.2

Table 7.
Nickel loading in batch with column or sandwich reactor configuration at 14 MPa, 200°C and 1 h.

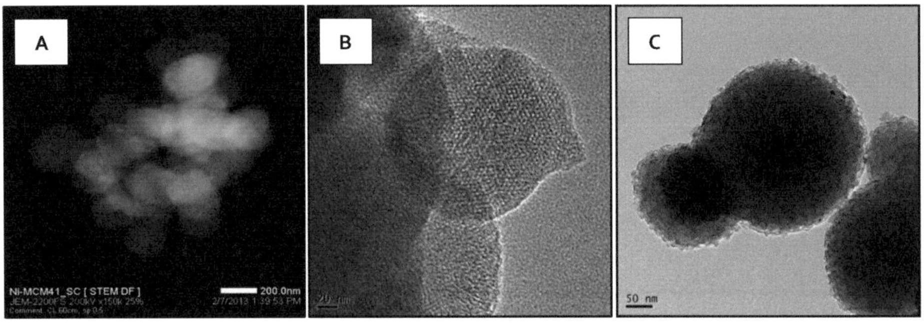

Figure 16.
TEM pictures of Ni-MCM-48 (A): BSE mode; (B and C) electronic image.

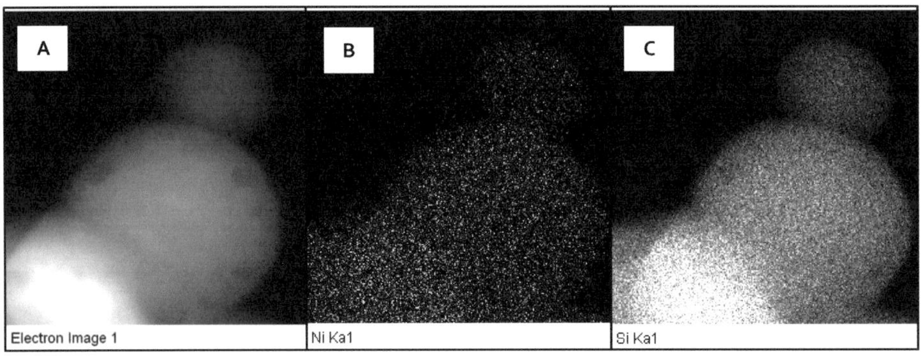

Figure 17.
Mapping of the SEM images obtained for the Ni-MCM-48 catalyst. (A) Electron image, (B) NiKα, and (C) SiKα spectra.

As occurred with the cobalt deposition, channels can be distinguished in **Figure 16**, which indicates that the high pressure of SCFD method does not modify the original structure of the support.

The distribution of the metal was observed by means of a mapping of the electron image (**Figure 17**). In these images, the homogenous distribution of the metal throughout the catalysts prepared by SCFD is observed, being perfectly distributed both surface and inside the support, without detecting any gradient of concentration or uncoated areas.

6.3 Ru-MCM-48 and Ru-C catalysts

6.3.1 Physical, chemical, and textural properties

Figure 18 shows the N_2 adsorption/desorption isotherms for the AC support and Ru-based catalysts. According to the IUPAC classification, the N_2 isotherms for AC and Ru-AC catalysts are classified as type I. Generally, the microporous solids such as activated carbons and molecular sieve zeolites give type I isotherms [171].

The quantitative data obtained from the isotherms are presented in **Table 8**. The surface area obtained for the carbonaceous support was 794 $m^2\,g^{-1}$, while the average diameter and pore volume were 2.1 nm and 0.77 $cm^3\,g^{-1}$, respectively. By incorporating 4.3 wt.% Ru with supercritical CO_2, the textural properties remain practically constant, being the surface area of 739 $m^2\,g^{-1}$, the pore size of 2.0 nm, and the volume of 0.70 $cm^3\,g^{-1}$. This behavior would indicate that ruthenium was successfully incorporated into the carbonaceous matrix, with high dispersion without collapsing the microporous structure.

On the other hand, when the ruthenium content was increased to 8 wt.%, the surface area decreased 30% compared to the initial AC. Also, pore size and volume decrease 23 and 12%, respectively.

In this way, it is verified that the supercritical fluid technology allowed to incorporate up to 8 wt.% of Ru in a microporous carbonaceous support in a satisfactory way without significantly altering the textural properties of the same.

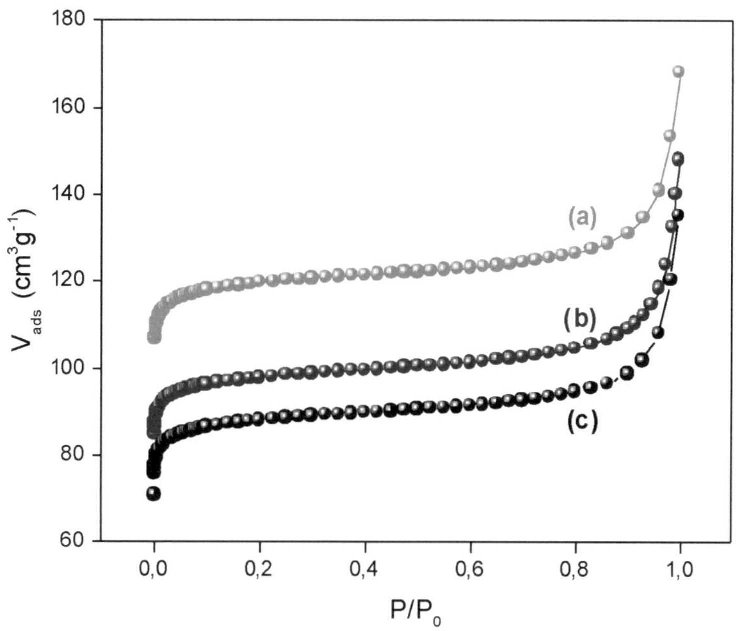

Figure 18.
Adsorption/desorption isotherms of N_2 at −196°C obtained for (a) AC support, (b) Ru(4.3)-AC, and (c) Ru(8)-AC catalysts.

Catalysts	% Ru[a]	S_{BET} ($m^2 \cdot g^{-1}$)	v_p ($cm^3 \cdot g^{-1}$)	t_p[b]
MCM-48	0	1641	0.90	3.6
Ru-MCM-48	3.0	977	0.45	1.4
AC	0	794	0.77	2.2
Ru(4.3)-AC	4.3	739	0.70	2.0
Ru(8)-AC	8	524	0.68	1.7

[a]*Determined by ICP (wt.%).*
[b]*Average pore size (nm).*

Table 8.
Chemical and textural properties of Ru-based catalysts.

Following the same line of research, when 4.7 wt.% Ru is incorporated into the MCM-48 mesoporous support, a 46% of decrease in the BET surface area is observed. Similar behavior shows the diameter and pore volume. This behavior was also observed with the incorporation of 3% Ni (see Section 6.2).

In this sense, it is observed that the modification of the textural properties of the synthesized catalysts depends on each particular case and factors like (i) the nature of the metal precursor used, (ii) the solubility of the precursor in sc-CO_2, and (iii) the incorporated metal content. Evidently, the optimized P and T conditions influence the solubility of the precursor but do not affect the structure of the catalytic supports.

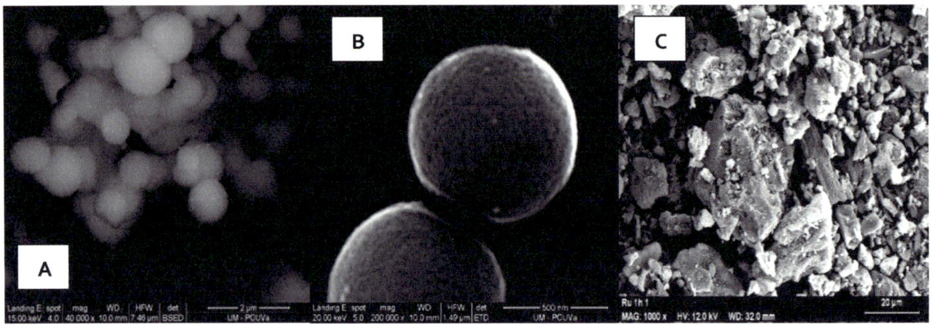

Figure 19.
SEM images of (A and B) Ru-MCM-48 and (C) Ru-C catalysts prepared by SCFD in continuous mode.

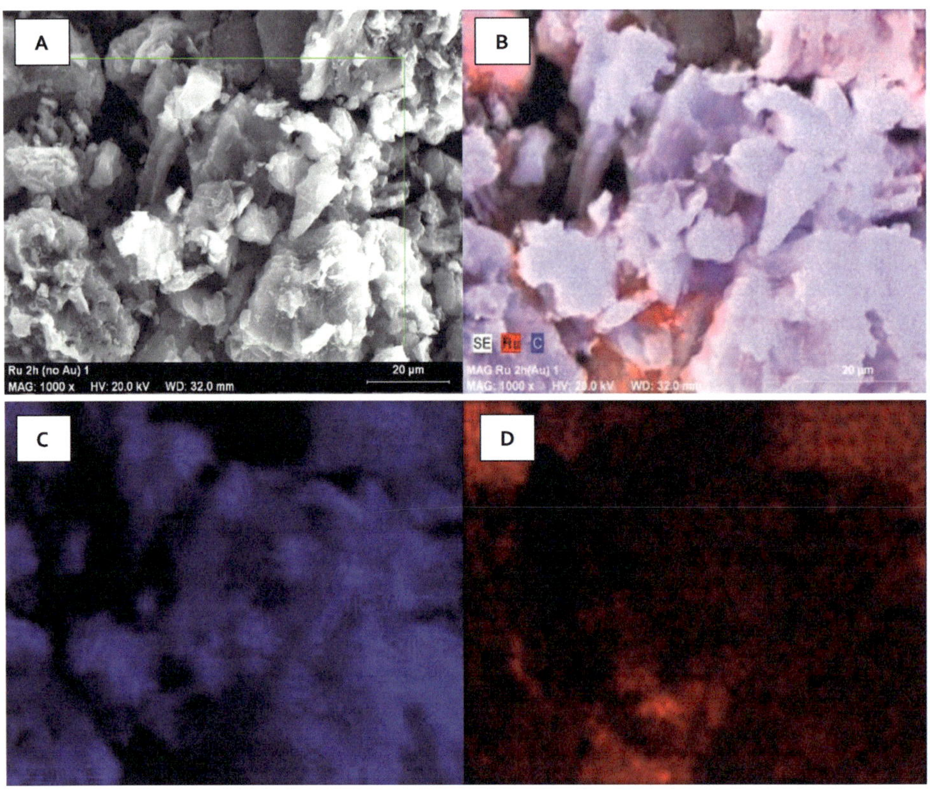

Figure 20.
Mapping of the SEM images obtained for the Ru-C catalyst. (A) Electron image, (B) SiKα, (C) CKα, and (D) RuKα spectra.

6.3.2 Morphology

Figure 19 presents the SEM images obtained after the addition of ruthenium on supports MCM-48 and activated carbon with sc-CO_2 in continuous conditions. Spherical particles with a size distribution between 200 and 500 nm were observed for MCM-48 support. Well-dispersed nanoparticles with an average diameter close to 10 nm were exposed on the MCM-48 surface (**Figure 19A** and **B**).

On the other hand, carbon active has a different morphology, irregularly shaped particles with very different sizes. **Figure 19C** presents the morphology corresponding to the Ru-C catalyst. From this SEM image, it is not possible to observe the deposited metal phase, so the mapping of the obtained electronic image was performed (**Figure 20**).

It is observed that the metal is completely deposited on the surface of the activated carbon with homogeneous distribution. Increasing the concentration of metal in the catalyst, it still maintains a uniform distribution, without finding gradient of concentration.

7. Conclusions

Supercritical CO_2 represents an optimal means to synthesize metallic nanoparticles (cobalt, nickel, or ruthenium) on substrates such as MCM-48, MCM-41, Al-MCM-41, and activated carbon.

It has important advantages over other conventional means of preparation, such as complete synthesis in a single stage, without the need for subsequent calcination or reduction. Furthermore, the synthesis process is simple, using CO_2 which is nontoxic, nonreactive, nonflammable, and inexpensive, with adequate pressure and temperature.

It is a process that can be performed in batch or continuous, optimizing the configuration of the reactor towards the sandwich arrangement of the support and organometallic precursor, thus obtaining a greater amount of catalyst in a single synthesis.

In this way, high metallic contents (close to 10%) are achieved with high dispersion and without significantly affecting the textural properties of the support.

Therefore, the effective deposition of different active phases on micro- and mesoporous substrates depends on each particular case and factors like the nature of the metal precursor used, the solubility of the precursor in sc-CO_2, and the incorporated metal content. Evidently, the optimized P and T conditions influence the solubility of the precursor but do not affect the structure of the catalytic supports.

Finally, reactive deposition with supercritical CO_2 represents an alternative for the reuse of CO_2 captured in other processes.

Acknowledgements

The authors acknowledge the financial support received from the Junta de Castilla y León, Spain (VA040u16), and from the Universidad Nacional del Litoral and CONICET, Argentina. They are also grateful to Santa Fe Agency of Science, Technology and Innovation (ASACTEI) for the financial support received through the Project 2017-00157.

Author details

Soledad Guadalupe Aspromonte[1*], Federico Andrés Piovano[1], Esther Alonso[2] and Alicia Viviana Boix[1]

1 INCAPE (FIQ, UNL-CONICET), Research Institute in Catalysis and Petrochemistry, Santa Fe, Argentina

2 High Pressure Processes Group, Department of Chemical Engineering and Environmental Technology, BioEcoUva, Research Institute on Bioeconomy, University of Valladolid, Valladolid, Spain

*Address all correspondence to: saspromonte@fiq.unl.edu.ar

References

[1] Poliakoff M, Fitzpatrick JM, Farre TR, Anastas PT. Green Chemistry: Science and politics of change. Science. 2002;**297**(5582):807-810. DOI: 10.1126/science.297.5582.807

[2] Anastas PT, Kirchhoff MM. Origins, currents status, and future challenges of Green Chemistry. Accounts of Chemical Research. 2002;**35**(9):686-694

[3] Yang Z, Pan W. Ionic liquids: Green solvents for nonaqueous biocatalysis. Enzyme and Microbial Technology. 2005;**37**(1):19-28

[4] Cho C-W, Jeon Y-C, Pham TPT, Vijayaraghavan K, Yun Y-S. The ecotoxicity of ionic liquids and traditional organic solvents on microalga *Selenastrum capricornutum*. Ecotoxicology and Environmental Safety. 2008;**71**(1):166-171

[5] Zhu P, Chen Y, Wang LY, Zhou M. Treatment of waste printed circuit board by green solvent using ionic liquid. Waste Management. 2012;**32**(10):1914-1918

[6] Nalawade SP, Picchioni F, Janssen LPBM. Supercritical carbon dioxide as a green solvent for processing polymer melts: Processing aspects and applications. Progress in Polymer Science. 2006;**31**(1):19-43

[7] Ramsey E, Sun Q, Zhang Z, Zhang C, Gou W. Mini-review: Green sustainable processes using supercritical fluid carbon dioxide. Journal of Environmental Sciences. 2009;**21**(6):720-726

[8] Machida H, Takesue M, Smith RL Jr. Green chemical processes with supercritical fluids: Properties, materials, separations and energy. The Journal of Supercritical Fluids. 2011;**60**:2-15

[9] McNaught AD, Wilkinson A. IUPAC Compendium of Chemical Terminology, the Gold Book. Oxford: Blackwell Scientific Publications; 2006

[10] Blackburn JM, Long DP, Cabañas A, Watkins JJ. Deposition of conformal copper and nickel films from supercritical carbon dioxide. Science. 2001;**294**:141-145

[11] Sotelo Sánchez JL, Ovejero Escudero G. Procesos con fluidos supercríticos. Anales de la Real Sociedad Española de Química. 2003;**4**:15-23

[12] Cansell F, Aymonier C. Review: Design of functional nanostructured materials using supercritical fluids. Journal of Supercritical Fluids. 2009;**47**:508-516

[13] Pai RA, Humayun R, Schulberg MT, Sebgupta A, Sun JN, Watkins JJ. Mesoporous silicates prepared using preorganized templates in supercritical fluids. Science. 2004;**303**:507-510

[14] Zhang J, Han B. Review: Supercritical CO_2-continuous microemulsions and compressed CO_2-expanded reverse microemulsions. Journal of Supercritical Fluids. 2007;**47**:179-186

[15] Reverchon E, Adami R. Review: Nanomaterials and supercritical fluids. Journal of Supercritical Fluids. 2006;**37**:1-22

[16] Knez Z, Markocic E, Leitgeb M, Primozic M, Knez Hrncic M, Skerget M. Industrial applications of supercritical fluids: A review. Energy. 2014;**77**:235-243

[17] Jung J, Perrut M. Particle design using supercritical fluids: Literature and patent survey. Journal of Supercritical Fluids. 2001;**20**:179-219

[18] Huang Z, Sun G-B, Chiew YC, Kawi S. Formation of ultrafine aspirin particles through rapid expansion of supercritical solution (RESS). Powder Technology. 2005;**160**(2):127-134

[19] Türk M, Bolten D. Formation of submicron poorly water-soluble drugs by rapid expansion of supercritical solution (RESS): Results for naproxen. The Journal of Supercritical Fluids. 2010;**55**(2):778-785

[20] Ovaskainen L, Chigome S, Birkin NA, Howdle SM, Torto N, Wagberg L, et al. Superhydrophobic polymeric coatings produced by rapid expansion of supercritical solutions combined with electrostatic deposition (RESS-ED). The Journal of Supercritical Fluids. 2014;**95**:610-617

[21] Baseri H, Lotfollahi MN. Formation of gemfibrozil with narrow particle size distribution via rapid expansion of supercritical solution process (RESS). Powder Technology. 2013;**235**:677-684

[22] Keshmiri K, Vatanara A, Tavakoli O, Manafi N. Production of ultrafine clobetasol propionate via rapid expansion of supercritical solution (RESS): Full factorial approach. The Journal of Supercritical Fluids. 2015;**101**:176-183

[23] Esfandiari N. Production of micro and nanoparticles of pharmaceutical by supercritical carbon dioxide. The Journal of Supercritical Fluids. 2015;**100**:129-141

[24] Wang H, Jiang H, Kuang L, Zhang M. Synthesis of highly dispersed MnOx-CeO_2 nanospheres by surfactant-assisted supercritical anti-solvent (SAS) technique. The important role of the surfactant. The Journal of Supercritical Fluids. 2014;**92**:84-92

[25] Adeli E. A comparative evaluation between utilizing SAS supercritical fluid technique and solvent evaporation method in preparation of azithromycin solid dispersions for dissolution rate enhancement. The Journal of Supercritical Fluids. 2014;**87**:9-21

[26] Cocero MJ, Martín Á, Mattea F, Varona S. Encapsulation and co-precipitation processes with supercritical fluids: Fundamentals and applications. The Journal of Supercritical Fluids. 2009;**47**(3):546-555

[27] Varona S, Kareth S, Martín Á, Cocero MJ. Formulation of lavandin essential oil with biopolymers by PGSS for application as biocide in ecological agriculture. The Journal of Supercritical Fluids. 2010;**54**(3):369-377

[28] Pestieau A, Krier F, Lebrun P, Brouwers A, Streel B, Evrard B. Optimization of a PGSS (particles from gas saturated solutions) process for a fenofibrate lipid-based solid dispersion formulation. International Journal of Pharmaceutics. 2015;**485**(1-2):295-305

[29] Bochon I, Kareth S, Kilzer A, Petermann M. Synthesis and powder generation of powder coatings using supercritical carbon dioxide. The Journal of Supercritical Fluids. 2015;**96**:324-333

[30] Ma CY, Liu JJ, Zhang Y, Wang XZ. Simulation for scale-up of a confined jet mixer for continuous hydrothermal flow synthesis of nanomaterials. The Journal of Supercritical Fluids. 2015;**98**:211-221

[31] Zhu K, Hu G. Supercritical hydrothermal synthesis of titanium dioxide nanostructures with controlled phase and morphology. The Journal of Supercritical Fluids. 2014;**94**:165-173

[32] Kawasaki S-I, Sue K, Ookawara R, Wakashima Y, Suzuki A, Hakuta Y, et al. Engineering study of continuous supercritical hydrothermal method using a T-shaped mixer: Experimental synthesis of NiO nanoparticles and CFD

simulation. The Journal of Supercritical Fluids. 2010;**54**(1):96-102

[33] Inoue M. Solvothermal synthesis of metal oxides. In: Somiya S, editor. Handbook of Advanced Ceramics. 2nd ed. Oxford: Academic Press; 2013. pp. 927-948

[34] Wang M, Chen C, Zhao B, Zeng Q, He D. Solvothermal synthesis of nanostructured TiO_2 photocatalyst in supercritical CO_2 fluids. Materials Letters. 2013;**109**:104-107

[35] Leng Y, Wang W, Zhang L, Zabihi F, Zhao Y. Fabrication and photocatalytical enhancement of ZnO-graphene hybrid using a continuous solvothermal technique. The Journal of Supercritical Fluids. 2014;**91**:61-67

[36] Malik MA, Wani MY, Hashim MA. Microemulsion method: A novel route to synthesize organic and inorganic nanomaterials: 1st nano update. Arabian Journal of Chemistry. 2012;**5**(4):397-417

[37] Blattner C, Zoumpanioti M, Kröner J, Schmeer G, Xenakis A, Kunz W. Biocatalysis using lipase encapsulated in microemulsion-based organogels in supercritical carbon dioxide. The Journal of Supercritical Fluids. 2006;**36**(3):182-193

[38] Shimizu R, Nibe A, Sawada K, Enokida Y, Yamamoto I. Preparation of hydrophobic platinum catalysts using a water-in-CO_2 microemulsion. The Journal of Supercritical Fluids. 2008;**44**(1):109-114

[39] Lu L, An X. Silver nanoparticles synthesis using H_2 as reducing agent in toluene-supercritical CO_2 microemulsion. The Journal of Supercritical Fluids. 2015;**99**:29-37

[40] Darr JA, Poliakoff M. New directions in inorganic and metal-organic coordination chemistry in supercritical fluids. Chemical Reviews. 1999;**99**(2):495-542

[41] Wenclawiak B, Bickmann F. Fluid and supercritical CO_2 as eluent in metal chelate chromatography. Analytical Chemistry. 1984;**319**:305-310

[42] Banister JA, George MW, Grubert S, Howdle SM, Jobling M, Johnson FPA, et al. Organometallic photochemistry in supercritical fluids: Reactions of cyclopentadienyl carbonyl and phosphine carbonyl complexes of manganese with dinitrogen. Journal of Organometallic Chemistry. 1994;**484**(1-2):129-135

[43] Lin Y, Smart NG, Wai CM. Supercritical fluid extraction and chromatography of metal chelates and organometallic compounds. TrAC Trends in Analytical Chemistry. 1995;**14**(3):123-133

[44] Bayona JM, Development of supercritical fluid extraction procedures for the determination of organotin compounds in sediment. In: Quevauviller EAM, Griepink B, editors. Techniques and Instrumentation in Analytical Chemistry. Netherlands: Elsevier BV; 1995. pp. 465-487

[45] Rathke JW, Klingler RJ, Krause TR. Propylene hydroformylation in supercritical carbon dioxide. Organometallics. 1991;**10**:1350-1355

[46] Dharmidhikari S, Abraham MA. Rhodium supported on activated carbon as a heterogeneous catalyst for hydroformylation of propylene in supercritical carbon dioxide. The Journal of Supercritical Fluids. 2000;**18**(1):1-10

[47] Bektesevic S, Kleman AM, Marteel-Parrish AE, Abraham MA. Hydroformylation in supercritical carbon dioxide: Catalysis and benign solvents. The Journal of Supercritical Fluids. 2006;**38**(2):232-241

[48] Koeken ACJ, de Bakker SJM, Costerus HM, van den Broeke LJP, Deelman B-J, Keurentjes JTF. Evaluation of pressure and correlation to reaction rates during homogeneously catalyzed hydroformylation in supercritical carbon dioxide. The Journal of Supercritical Fluids. 2008;**46**(1):47-56

[49] Laintz KE, Wai CM, Yonker CR, Smith RD. Extraction of metal ions from liquid and solid materials by supercritical carbon dioxide. Analytical Chemistry. 1992;**64**:28-75

[50] Liang M-T, Liang R-C, Lin C-H, Hsu P-J, Wu L-Y, Chen H-F, et al. Metal extraction of a spiked solid with supercritical carbon dioxide. The Journal of Supercritical Fluids. 2013;**79**:324-329

[51] Erkey C. Supercritical carbon dioxide extraction of metals from aqueous solutions: A review. The Journal of Supercritical Fluids. 2000;**17**(3):259-287

[52] Reverchon E, Adami R. Nanomaterials and supercritical fluids. The Journal of Supercritical Fluids. 2006;**37**(1):1-22

[53] Shah PS, Husain S, Johnston KP, Korgel BA. Nanocrystal arrested precipitation in supercritical carbon dioxide. Journal of Physical Chemistry. 2001;**105**:9433-9440

[54] Mezian IMJ, Pathak P, Beacham F, Allard LF, Sun Y-P. Nanoparticle formation in rapid expansion of water-in-supercritical carbon dioxide microemulsion into liquid solution. The Journal of Supercritical Fluids. 2005;**34**(1):91-97

[55] Haldorai Y, Shim J-J, Lim KT. Synthesis of polymer-inorganic filler nanocomposites in supercritical CO_2. The Journal of Supercritical Fluids. 2012;**71**:45-63

[56] Aymonier C, Loppinet-Serani A, Reverón H, Garrabos Y, Cansell F. Review of supercritical fluids in inorganic materials science. The Journal of Supercritical Fluids. 2006;**38**(2):242-251

[57] Higginbotham CL, Yons JGL, Kennedy JE. Polymer processing using supercritical fluids. In: Thomas S, Weimin Y, editors. Advances in Polymer Processing. Sawston, Cambridge: Woodhead Publishing; 2009. pp. 384-401

[58] Hussain YA, Grant CS. Ibuprofen impregnation into submicron polymeric films in supercritical carbon dioxide. The Journal of Supercritical Fluids. 2012;**71**:127-135

[59] Cooper AI, Kazarian SG, Poliakoff M. Supercritical fluid impregnation of polyethylene films, a new approach to studying equilibria in matrices; the hydrogen bonding of fluoroalcohols to (η^5-C_5Me_5)$Ir(CO)_2$ and the effect on C·H activation. Chemical Physics Letters. 1993;**206**(1-4):175-180

[60] Tomasko DL, Han X, Liu D, Gao W. Supercritical fluid applications in polymer nanocomposites. Current Opinion in Solid State and Materials Science. 2003;7(4-5):407-412

[61] Watkins JJ, McCarthy TJ. Polymer/metal nanocomposite synthesis in supercritical CO_2. Chemistry of Materials. 1995;7:1991-1994

[62] Bozbağ SE, Erkey C. Supercritical deposition: Current status and perspectives for the preparation of supported metal nanostructures. The Journal of Supercritical Fluids. 2015;**96**:298-312

[63] Erkey C. Preparation of metallic supported nanoparticles and films using supercritical fluid deposition. The Journal of Supercritical Fluids. 2009;**47**(3):517-522

[64] Caputo G, De Marco I, Reverchon E. Silica aerogel-metal composites produced by supercritical adsorption. The Journal of Supercritical Fluids. 2010;**54**(2):243-249

[65] Aspromonte SG, Sastre A, Boix AV, Cocero MJ, Alonso E. Cobalt oxide nanoparticles on mesoporous MCM-41 and Al-MCM-41 by supercritical CO_2 deposition. Microporous and Mesoporous Materials. 2012;**148**(1):53-61

[66] Alonso E, Montequi I, Lucas S, Cocero MJ. Synthesis of titanium oxide particles in supercritical CO_2: Effect of operational variables in the characteristics of the final product. The Journal of Supercritical Fluids. 2007;**39**(3):453-461

[67] Alonso E, Montequi I, Cocero MJ. Effect of synthesis conditions on photocatalytic activity of TiO_2 powders synthesized in supercritical CO_2. The Journal of Supercritical Fluids. 2009;**49**(2):233-238

[68] Bozbag SE, Sanli D, Erkey C. Synthesis of nanostructures materials using supercritical CO_2: Physical transformations. Journal of Material Sciences. 2012;**47**:2995-3025

[69] Bozbag SE, Sanli D, Erkey C. Synthesis of nanostructures materials using supercritical CO_2: Chemical transformations. Journal of Material Sciences. 2012;**47**:3469-3492

[70] Sinha Ray S. Applications, environmental impact and future development of environmentally friendly polymer nanocomposites (EFPNCs). In: Ray SS, editor. Environmentally Friendly Polymer Nanocomposites. Sawston, Cambridge: Woodhead Publishing; 2013. pp. 467-477

[71] Fitzgerald G, DeJoannis J, Meunier M. Multiscale modeling of nanomaterials: Recent developments and future prospects. In: Zhang VKT, editor. Modeling, Characterization, and Production of Nanomaterials. Sawston, Cambridge: Woodhead Publishing; 2015. pp. 3-53

[72] Pereira SR, Coelho MC. Can nanomaterials be a solution for application on alternative vehicles?—A review paper on life cycle assessment and risk analysis. International Journal of Hydrogen Energy. 2015;**40**(14):4969-4979

[73] Jiang K, Pinchuk AO. Noble metal nanomaterials: Synthetic routes, fundamental properties, and promising applications. In: Solid State Physics. Cambridge: Academic Press; 2015

[74] Wu R, Zhou K, Yue CY, Wei J, Pan Y. Recent progress in synthesis, properties and potential applications of SiC nanomaterials. Progress in Materials Science. 2015;**72**:1-60

[75] Zhang Y, Erkey C. Preparation of supported metallic nanoparticles using supercritical fluids: A review. The Journal of Supercritical Fluids. 2006;**38**(2):252-267

[76] Yen CH, Lin HW, Phan TD, Tan CS. Chemical fluid deposition of monometallic and bimetallic nanoparticles on ordered mesoporous silica as hydrogenation catalysts. Journal of Nanoscience and Nanotechnology. 2011;**11**(3):2465-2469

[77] Cangül B, Zhang LC, Aindow M, Erkey C. Preparation of carbon black supported Pd, Pt and Pd-Pt nanoparticles using supercritical CO_2 deposition. The Journal of Supercritical Fluids. 2009;**50**(1):82-90

[78] Garrido E, Aymonier C, Roiban L, Ersen O, Labrugère C, Gaillard P, et al. Noble metals supported on carbon nanotubes using supercritical fluids for the preparation of composite materials:

A look at the interface. The Journal of Supercritical Fluids. 2015;**101**:110-116

[79] Sanli D, Erkey C. Effect of polymer molecular weight and deposition temperature on the properties of silica aerogel/hydroxy-terminated poly(dimethylsiloxane) nanocomposites prepared by reactive supercritical deposition. The Journal of Supercritical Fluids. 2015;5(22):11708-11717

[80] Wolff S, Crone M, Muller T, Enders M, Bräse S, Türk M. Preparation of supported Pt nanoparticles by supercritical fluid reactive deposition: Influence of precursor, substrate and pressure on product properties. The Journal of Supercritical Fluids. 2014;**95**:588-596

[81] Müller S, Türk M. Production of supported gold and gold-silver nanoparticles by supercritical fluid reactive deposition: Effect of substrate properties. The Journal of Supercritical Fluids. 2015;**96**:287-297

[82] Kameo A, Yoshimura T, Esumi K. Preparation of noble metal nanoparticles in supercritical carbon dioxide. Colloids and Surfaces A: Physicochemical and Engineering Aspects. 2003;**215**:181-189

[83] Erkey C. Review: Preparation of metallic supported nanoparticles and films using supercritical fluid deposition. Journal of Supercritical Fluids. 2009;**47**:517-522

[84] Ciesla U, Schüth F. A review—Ordered mesoporous materials. Microporous and Mesoporous Materials. 1999;**27**:131-149

[85] Melero JA, Iglesias J, Moreno J. Chapter 12: Advanced metal-containing mesostructured silicas for novel catalytic applications. In: Mesoporous Materials: Properties, Preparation and Applications. NY: Nova Science Publishers; 2009

[86] Chen CY, Burkett SL, Li HX, Davis ME. Studies on mesoporous materials. II. Synthesis mechanism of MCM-41. Microporous Materials. 1993;**2**:21-34

[87] Beck JS, Vartuli JC. Recent advances in the synthesis, characterization and applications of mesoporous molecular sieves. Current Opinion in Solid State & Materials Science. 1996;**1**:76-87

[88] Zhou W, Klinowski J. The mechanism of channel formation in the mesoporous molecular sieve MCM-41. Chemical Physics Letters. 1998;**292**:207-212

[89] Taguchi A, Schüth F. Ordered mesoporous materials in catalysis—Review. Microporous and Mesoporous Materials. 2005;**77**:1-45

[90] Trong On D, Desplantier-Giscard D, Danumah C, Kaliaguine S. Perspectives in catalytic applications of mesostructured materials—A review. Applied Catalysis A: General. 2001;**222**:299-357

[91] Luechinger M, Frunz L, Pirngruber GD, Prins R. A mechanistic explanation of the formation of high quality MCM-41 with high hydrothermal stability. Microporous and Mesoporous Materials. 2003;**64**:203-211

[92] Landau MV, Dafa E, Kaliya ML, Sen T, Herskowitz M. Mesoporous alumina catalytic material prepared by grafting wide-pore MCM-41 with an alumina multilayer. Microporous and Mesoporous Materials. 2001;**49**:65-81

[93] Passos AR, Martins L, Pulcinelli SH, Santilli CV, Briois V. Effect of the balance between Co(II) and Co(0) oxidation states on the catalytic activity of cobalt catalysts for ethanol steam reforming. Catalysis Today. 2014;**229**:88-94

[94] Pairojpiriyakul T, Croiset E, Kiatkittipong W, Kiatkittipong K, Arpornwichanop A, Assabumrungrat S. Hydrogen production from catalytic supercritical water reforming of glycerol with cobalt-based catalysts. International Journal of Hydrogen Energy. 2013;**38**(11):4368-4379

[95] Chiou JYZ, Lee C-L, Ho K-F, Huang H-H, Yu S-W, Wang C-W. Catalytic performance of Pt-promoted cobalt-based catalysts for the steam reforming of ethanol. International Journal of Hydrogen Energy. 2014;**39**(11):5653-5662

[96] Garbarino G, Riani P, Lucchini MA, Canepa F, Kawale S, Busca G. Cobalt-based nanoparticles as catalysts for low temperature hydrogen production by ethanol steam reforming. International Journal of Hydrogen Energy. 2013;**38**(1):82-91

[97] Papadopoulou E, Delimaris D, Denis A, Machocki A, Ioannides T. Alcohol reforming on cobalt-based catalysts prepared from organic salt precursors. International Journal of Hydrogen Energy. 2012;**37**(21):16375-16381

[98] Llorca J. Efficient production of hydrogen over supported cobalt catalysts from ethanol steam reforming. Journal of Catalysis. 2002;**209**(2):306-317

[99] Busca G, Costantino U, Montanari T, Ramis G, Resini C, Sisani M. Nickel versus cobalt catalysts for hydrogen production by ethanol steam reforming: Ni-Co-Zn-Al catalysts from hydrotalcite-like precursors. International Journal of Hydrogen Energy. 2010;**35**(11):5356-5366

[100] Chen L, Choong CKS, Zhong Z, Huang L, Wang Z, Lin J. Support and alloy effects on activity and product selectivity for ethanol steam reforming over supported nickel cobalt catalysts. International Journal of Hydrogen Energy. 2012;**37**(21):16321-16332

[101] Resini C, Concepción Herrera Delgado M, Presto S, Alemany LJ, Riani P, Marazza R, et al. Yttria-stabilized zirconia (YSZ) supported Ni-Co alloys (precursor of SOFC anodes) as catalysts for the steam reforming of ethanol. International Journal of Hydrogen Energy. 2008;**33**(14):3728-3735

[102] Pereira EB, de la Piscina PR, Homs N. Efficient hydrogen production from ethanol and glycerol by vapour-phase reforming processes with new cobalt-based catalysts. Bioresource Technology. 2011;**102**(3):3419-3423

[103] de la Peña O'Shea VA, Nafria R, Ramírez de la Piscina P, Homs N. Development of robust Co-based catalysts for the selective H_2-production by ethanol steam-reforming. The Fe-promoter effect. International Journal of Hydrogen Energy. 2008;**33**(13):3601-3606

[104] Casanovas A, Roig M, de Leitenburg C, Trovarelli A, Llorca J. Ethanol steam reforming and water gas shift over Co/ZnO catalytic honeycombs doped with Fe, Ni, Cu, Cr and Na. International Journal of Hydrogen Energy. 2010;**35**(15):7690-7698

[105] Xiong H, Motchelaho MAM, Moyo M, Jewell LL, Coville NJ. Cobalt catalysts supported on a micro-coil carbon in Fischer-Tropsch synthesis: A comparison with CNTs and CNFs. Catalysis Today. 2013;**214**:50-60

[106] Xiong H, Zhang Y, Liew K, Li J. Ruthenium promotion of Co/SBA-15 catalysts with high cobalt loading for Fischer-Tropsch synthesis. Fuel Processing Technology. 2009;**90**(2):237-246

[107] Griboval-Constant A, Butel A, Ordomsky VV, Chernavskii PA,

Khodakov AY. Cobalt and iron species in alumina supported bimetallic catalysts for Fischer-Tropsch reaction. Applied Catalysis A: General. 2014;**481**:116-126

[108] Martínez AN, López C, Márquez F, Díaz I. Fischer-Tropsch synthesis of hydrocarbons over mesoporous Co/SBA-15 catalysts: The influence of metal loading, cobalt precursor, and promoters. Journal of Catalysis. 2003;**220**(2):486-499

[109] Wu H, Yang Y, Suo H, Qing M, Yan L, Wu B, et al. Effect of TiO_2 promotion on the structure and performance of silica-supported cobalt-based catalysts for Fischer-Tropsch synthesis. Journal of Molecular Catalysis A: Chemical. 2014;**390**:52-62

[110] Davari M, Karimi S, Tavasoli A, Karimi A. Enhancement of activity, selectivity and stability of CNTs-supported cobalt catalyst in Fischer-Tropsch via CNTs functionalization. Applied Catalysis A: General. 2014;**485**:133-142

[111] Parnian MJ, Taheri Najafabadi A, Mortazavi Y, Khodadadi AA, Nazzari I. Ru promoted cobalt catalyst on γ-Al_2O_3: Influence of different catalyst preparation method and Ru loadings on Fischer-Tropsch reaction and kinetics. Applied Surface Science. 2014;**313**:183-195

[112] Cook KM, Perez HD, Bartholomew CH, Hecker WC. Effect of promoter deposition order on platinum-, ruthenium-, or rhenium-promoted cobalt Fischer-Tropsch catalysts. Applied Catalysis A: General. 2014;**482**:275-286

[113] Karolewska M, Truszkiewicz E, Mierzwa B, Kępiński L, Raróg-Pilecka W. Ammonia synthesis over cobalt catalysts doped with cerium and barium. Effect of the ceria loading. Applied Catalysis A: General. 2012;**445-446**:280-286

[114] Karolewska M, Truszkiewicz E, Wściseł M, Mierzwa B, Kępiński L, Raróg-Pilecka W. Ammonia synthesis over a Ba and Ce-promoted carbon-supported cobalt catalyst. Effect of the cerium addition and preparation procedure. Journal of Catalysis. 2013;**303**:130-134

[115] Campa MC, Luisetto I, Pietrogiacomi D, Indovina V. The catalytic activity of cobalt-exchanged mordenites for the abatement of NO with CH_4 in the presence of excess O_2. Applied Catalysis B: Environmental. 2003;**46**(3):511-522

[116] Campa MC. Catalytic activity of Co-ZSM-5 for the abatement of NOx, with methane in the presence of oxygen. Applied Catalysis B: Environmental. 1996;**8**:315-331

[117] Yan Y, Wang L, Zhang H. Catalytic combustion of volatile organic compounds over Co/ZSM-5 coated on stainless steel fibers. Chemical Engineering Journal. 2014;**255**:195-204

[118] Ulla MA, Lombardo E, Daniell W, Knözinger H. Catalytic combustion of methane on Co/MgO: Characterisation of active cobalt sites. Applied Catalysis B: Environmental. 2001;**29**:217-229

[119] Jiang Z, Yu J, Cheng J, Xiao T, Jones MO, Hao Z, et al. Catalytic combustion of methane over mixed oxides derived from Co-Mg/Al ternary hydrotalcites. Fuel Processing Technology. 2010;**91**(1):97-102

[120] Hunde ET, Watkins JJ. Reactive deposition of cobalt and nickel films from their metallocenes in supercritical carbon dioxide solution. Chemistry of Materials. 2004;**16**(3):498-503

[121] Voorhoeve RJH, Stuiver JCM. The mechanism of the hydrogenation of cyclohexene and benzene on nickel-tungsten sulfide catalysts. Journal of Catalysis. 1971;**23**(2):243-252

[122] Bang Y, Han SJ, Seo JG, Youn MH, Song JH, Song IK. Hydrogen production by steam reforming of liquefied natural gas (LNG) over ordered mesoporous nickel-alumina catalyst. International Journal of Hydrogen Energy. 2012;**37**(23):17967-17977

[123] Majewski AJ, Wood J, Bujalski W. Nickel-silica core@shell catalyst for methane reforming. International Journal of Hydrogen Energy. 2013;**38**(34):14531-14541

[124] Han SJ, Bang Y, Yoo J, Seo JG, Song IK. Hydrogen production by steam reforming of ethanol over mesoporous Ni-Al_2O_3-ZrO_2 xerogel catalysts: Effect of nickel content. International Journal of Hydrogen Energy. 2013;**38**(20):8285-8292

[125] Bimbela F, Oliva M, Ruiz J, García L, Arauzo J. Hydrogen production via catalytic steam reforming of the aqueous fraction of bio-oil using nickel-based coprecipitated catalysts. International Journal of Hydrogen Energy. 2013;**38**(34):14476-14487

[126] Morlanés N. Reaction mechanism of naphtha steam reforming on nickel-based catalysts, and FTIR spectroscopy with CO adsorption to elucidate real active sites. International Journal of Hydrogen Energy. 2013;**38**(9):3588-3596

[127] Nieva MA, Villaverde MM, Monzón A, Garetto TF, Marchi AJ. Steam-methane reforming at low temperature on nickel-based catalysts. Chemical Engineering Journal. 2014;**235**:158-166

[128] Hashemnejad SM, Parvari M. Deactivation and regeneration of nickel-based catalysts for steam-methane reforming. Chinese Journal of Catalysis. 2011;**32**(1):273-279

[129] Lua AC, Wang HY. Decomposition of methane over unsupported porous nickel and alloy catalyst. Applied Catalysis B: Environmental. 2013;**132-133**:469-478

[130] Abelló S, Berrueco C, Montané D. High-loaded nickel-alumina catalyst for direct CO_2 hydrogenation into synthetic natural gas (SNG). Fuel. 2013;**113**:598-609

[131] Lovell E, Jiang Y, Scott J, Wang F, Suhardja Y, Chen M, et al. CO_2 reforming of methane over MCM-41-supported nickel catalysts: Altering support acidity by one-pot synthesis at room temperature. Applied Catalysis A: General. 2014;**473**:51-58

[132] Zhang L, Li L, Zhang Y, Zhao Y, Li J. Nickel catalysts supported on MgO with different specific surface area for carbon dioxide reforming of methane. Journal of Energy Chemistry. 2014;**23**(1):66-72

[133] Tao K, Shi L, Ma Q, Wang D, Zeng C, Kong C, et al. Methane reforming with carbon dioxide over mesoporous nickel-alumina composite catalyst. Chemical Engineering Journal. 2013;**221**:25-31

[134] Fukuhara C, Hyodo R, Yamamoto K, Masuda K, Watanabe R. A novel nickel-based catalyst for methane dry reforming: A metal honeycomb-type catalyst prepared by sol-gel method and electroless plating. Applied Catalysis A: General. 2013;**468**:18-25

[135] Ranjbar A, Rezaei M. Preparation of nickel catalysts supported on $CaO.2Al_2O_3$ for methane reforming with carbon dioxide. International Journal of Hydrogen Energy. 2012;**37**(8):6356-6362

[136] Bereketidou OA, Goula MA. Biogas reforming for syngas production over nickel supported on ceria-alumina catalysts. Catalysis Today. 2012;**195**(1):93-100

[137] Huang L, Zhang F, Chen R, Hsu AT. Manganese-promoted

nickel/alumina catalysts for hydrogen production via auto-thermal reforming of ethanol. International Journal of Hydrogen Energy. 2012;**37**(21):15908-15913

[138] Chan FL, Tanksale A. Review of recent developments in Ni-based catalysts for biomass gasification. Renewable and Sustainable Energy Reviews. 2014;**38**:428-438

[139] Phuhiran C, Takarada T, Chaiklangmuang S. Hydrogen-rich gas from catalytic steam gasification of eucalyptus using nickel-loaded Thai brown coal char catalyst. International Journal of Hydrogen Energy. 2014;**39**(8):3649-3656

[140] Sabatier P, Senderens JB. New synthesis of methane. Comptes Rendus de l'Académie des Sciences. 1902;**134**:514-524

[141] Rahmani S, Rezaei M, Meshkani F. Preparation of promoted nickel catalysts supported on mesoporous nanocrystalline gamma alumina for carbon dioxide methanation reaction. Journal of Industrial and Engineering Chemistry. 2014;**38**(1):118-126

[142] Rahmani S, Rezaei M, Meshkani F. Preparation of highly active nickel catalysts supported on mesoporous nanocrystalline γ-Al_2O_3 for CO_2 methanation. Journal of Industrial and Engineering Chemistry. 2014;**20**(4):1346-1352

[143] Azadi P, Khan S, Strobel F, Azadi F, Farnood R. Hydrogen production from cellulose, lignin, bark and model carbohydrates in supercritical water using nickel and ruthenium catalysts. Applied Catalysis B: Environmental. 2012;**117-118**:330-338

[144] Upare PP, Jeong M-G, Hwang YK, Kim DH, Kim YD, Hwang DW, et al. Nickel-promoted copper-silica nanocomposite catalysts for hydrogenation of levulinic acid to lactones using formic acid as a hydrogen feeder. Applied Catalysis A: General. 2015;**491**:127-135

[145] Peng Q, Spagnola JC, Parsons GN. Self-catalyzed hydrogenolysis of nickelocene: Functional metal coating of three-dimensional nanosystems at low temperature. Journal of the Electrochemical Society. 2008;**155**(9):D580-D582

[146] Bozbag SE, Zhang LC, Aindow M, Erkey C. Carbon aerogel supported nickel nanoparticles and nanorods using supercritical deposition. The Journal of Supercritical Fluids. 2012;**66**:265-273

[147] Taylor AD, DiLeo GJ, Sun K. Hydrogen production and performance of nickel-based catalysts synthesized using supercritical fluids for the gasification of biomass. Applied Catalysis B: Environmental. 2009;**93**(1-2):126-133

[148] Babuněk M, Šimůnek O, Hošek J, Rybáčková M, Cvačka J, Březinová A, et al. Heavy fluorous phosphine-free ruthenium catalysts for alkene metathesis. Journal of Fluorine Chemistry. 2014;**161**:66-75

[149] Liu G, Zhang H, Zhao X, Wang J. Efficient ruthenium metathesis catalysts containing carborane ligands. Journal of Organometallic Chemistry. 2014;**749**:13-17

[150] Gao Y, Jaenicke S, Chuah G-K. Highly efficient transfer hydrogenation of aldehydes and ketones using potassium formate over AlO(OH)-entrapped ruthenium catalysts. Applied Catalysis A: General. 2014;**484**:51-58

[151] Durap F, Aydemir M, Baysal A, Elma D, Ak B, Turgut Y. A new efficient bis(phosphinite)-ruthenium(II) catalyst system for the asymmetric transfer

hydrogenation of aromatic ketones. Inorganica Chimica Acta. 2014;**411**:77-82

[152] Wang Z, Lin J, Wang R, Wei K. Ammonia synthesis over ruthenium catalyst supported on perovskite type $BaTiO_3$. Catalysis Communications. 2013;**32**:11-14

[153] Li Y, Pan C, Han W, Chai H, Liu H. An efficient route for the preparation of activated carbon supported ruthenium catalysts with high performance for ammonia synthesis. Catalysis Today. 2011;**174**(1):97-105

[154] Akbayrak S, Kaya M, Volkan M, Özkar S. Ruthenium(0) nanoparticles supported on magnetic silica coated cobalt ferrite: Reusable catalyst in hydrogen generation from the hydrolysis of ammonia-borane. Journal of Molecular Catalysis A: Chemical. 2014;**394**:253-261

[155] Akbayrak S, Tanyıldızı S, Morkan I, Özkar S. Ruthenium(0) nanoparticles supported on nanotitania as highly active and reusable catalyst in hydrogen generation from the hydrolysis of ammonia borane. International Journal of Hydrogen Energy. 2014;**39**(18):9628-9637

[156] Ma Y, Huang Y, Cheng Y, Wang L, Li X. Biosynthesized ruthenium nanoparticles supported on carbon nanotubes as efficient catalysts for hydrogenation of benzene to cyclohexane: An eco-friendly and economical bioreduction method. Applied Catalysis A: General. 2014;**484**:154-160

[157] Liao H, Ouyang D, Zhang J, Xiao Y, Liu P, Hao F, et al. Benzene hydrogenation over oxide-modified MCM-41 supported ruthenium-lanthanum catalyst: The influence of zirconia crystal form and surface hydrophilicity. Chemical Engineering Journal. 2014;**243**:207-216

[158] Arena BJ. Deactivation of ruthenium catalysts in continuous glucose hydrogenation. Applied Catalysis A: General. 1992;**87**:219-229

[159] Mishra DK, Dabbawala AA, Park JJ, Jhung SH, Hwang J-S. Selective hydrogenation of d-glucose to d-sorbitol over HY zeolite supported ruthenium nanoparticles catalysts. Catalysis Today. 2014;**232**:99-107

[160] Hao C, Wang S, Li M, Kang L, Ma X. Hydrogenation of CO_2 to formic acid on supported ruthenium catalysts. Catalysis Today. 2011;**160**(1):184-190

[161] Yen CH, Lin HW, Tan C-S. Hydrogenation of bisphenol A—Using a mesoporous silica based nano ruthenium catalyst Ru/MCM-41 and water as the solvent. Catalysis Today. 2011;**174**(1):121-126

[162] Kim D, Lee HBR, Yoon J, Kim H. Ru nanodot synthesis using CO_2 supercritical fluid deposition. Journal of Physics and Chemistry of Solids. 2013;**74**(5):664-667

[163] Zhang Y, Jiang H, Wang Y, Zhang M. Synthesis of highly dispersed ruthenium nanoparticles supported on activated carbon via supercritical fluid deposition. Industrial & Engineering Chemistry Research. 2014;**53**(15):6380-6387

[164] Xu S, Li L, Li R, Su Y. Preparation of Ru/C catalysts by supercritical CO_2 deposition. Xiyou Jinshu Cailiao Yu Gongcheng/Rare Metal Materials and Engineering. 2011;**40**(12):2142-2146

[165] Zhao Y, Zhang J, Song J, Li J, Liu J, Wu T, et al. Ru nanoparticles immobilized on metal-organic framework nanorods by supercritical CO_2-methanol solution: Highly efficient catalyst. Green Chemistry. 2011;**13**(8):2078-2082

[166] Szègedi A, Kónya Z, Méhn D, Solymár E, Pál-Borbély G, Horváth ZE,

et al. Spherical mesoporous MCM-41 materials containing transition metals: Synthesis and characterization. Applied Catalysis A: General. 2004;**272**:257-266

[167] Sobczak I, Ziolek M, Nowacka M. Preparation and characterization of Pt containing NbMCM-41 mesoporous molecular sieves addressed to catalytic NO reduction by hydrocarbons. Microporous and Mesoporous Materials. 2005;**78**:103-116

[168] Shanbhag GV, Joseph T, Halligudi SB. Copper(II) ion exchanged AlSBA-15: A versatile catalyst for intermolecular hydroamination of terminal alkynes with aromatic amines. Journal of Catalysis. 2007;**250**:274-282

[169] Brunauer S, Deming LS, Deming WE, Teller E. On a theory of the Van der Waals adsorption of gases. Journal of the American Chemical Society. 1940;**62**:1723-1732

[170] Morey M, Davidson A, Eckert H, Stucky G. Pseudotetrahedral O_3/2VO centers immobilized on the walls of a mesoporous, cubic MCM-48 support: Preparation, characterization and reactivity toward water as investigated by 51V NMR and UV-Vis spectroscopies. Chemistry of Materials. 1996;**8**:486-492

[171] Sing KS, Everett DH, Haul RAW, Moscou L, Pierotti RA, Rouquerol J, et al. Reporting physisorption data for gas/solid systems with special reference to the determination of surface area and porosity. Pure and Applied Chemistry. 1985;**57**:603-619

Chapter 2

Sulfided NiMo/*Clinoptilolite* Catalysts for Selective Sulfur Removal from Naphtha Stream without Olefin Hydrogenation

Cristina Farías Rosales, Rut Guil-López, Marisol Faraldos, Rafael Maya Yescas, Trino Armando Zepeda, Barbara Pawelec and Rafael Huirache-Acuña

Abstract

The natural *clinoptilolite* zeolite has been modified by acid leaching with HNO_3 in order to obtain economic material for supporting NiMoS hydrotreating catalysts. The most optimized zeolite material was obtained by leaching with HNO_3 at 80°C during 24 h. The bimetallic NiMo catalysts prepared by wet impregnation of a zeolite support, followed by calcination and sulfidation, were characterized by several physico-chemical techniques and tested in the hydrodesulfurization (HDS) of 3-methyl-thiophene (3-MT) model feed at atmospheric H_2 pressure and T = 280°C. For all catalysts, the 3-MT transformation mainly occurs via direct desulfurization reaction route being diminished the catalyst hydrogenation function. This was linked with the formation of highly stacked layers of MoS_2 particles having a low amount of "brim sites," as demonstrated by HRTEM. The cause of the best performance of Ni-$Mo_{(H)}$/Z-1 sulfide catalyst in the HDS of 3-MT can be the presence of K^+ impurities on the support surface which forces the formation of highly stacked layers of MoS_2 particles.

Keywords: hydrodesulfurization, acid leaching, *clinoptilolite*, zeolite, naphta

1. Introduction

The use of suitable natural zeolites for supporting heterogeneous catalysts is economically attractive option in many countries possessing abundance of natural zeolites [1]. Among these zeolites, the *clinoptilolite* is of considerable interest due to its effective high cation exchange capacity, unique crystal and pore structure, chemical stability in corrosive media, and thermostability [2]. *Clinoptilolite* has a cage-like structure consisting of SiO_4 and AlO_4 tetrahedral units joined by oxygen atoms. The native charges of the AlO_4 units are balanced by Mg, Ca, Na, K, and/or Fe cations. Moreover, its textural and acid properties should be improved by elimination of the extraframework cations, for example, by chemical leaching [3–5].

Clinoptilolite is crystalline aluminosilicate mineral which exhibits large number of acid sites distributed through the network of channels and cavities consisting of AlO_4 and SiO_2 tetrahedral units joined by shared oxygen atom [3]. It is a sheet-like structural organization that contains open 10-membered rings (7.5 Å × 3.1 Å) alternated with eight-membered rings (4.6 Å × 3.6 Å, 4.7 Å × 2.8 Å). Those rings are stacking together from sheet to sheet to form channels throughout the crystal structure [3]. The AlO_4 units exhibit negative charge which is compensated by the presence of undesired cations, such as Mg^{2+}, Ca^{2+}, Na^+, K^+, and/or Fe^{3+}. The number of acid sites and their strength can be controlled by zeolite dealumination, isomorphous substitution of atoms with tetrahedral or via ion exchange methods. The hydroxyl group generation (formation of Brønsted acid sites) can be achieved also by hydrolysis of a zeolite possessing multivalent cations, or by decomposition of the NH_4^+ ions into the zeolite. In this sense, the study by Arcoya et al. [4] demonstrated that the chemical treatment with NH_4Cl or HCl solutions opened the channels and increased acidity and thermal stability of *clinoptilolite*. On the contrary to HCl treatment, it was reported that HNO_3-treatment led to morphology changes of the raw *clinoptilolite* from the lead-like to the needle-like, an increase of specific surface area and decrease of crystallinity [5].

Recently, there is growing interest in the use of natural zeolites such as *clinoptilolite* as adsorbents [6–11], catalysts [12–17] or for supporting heterogeneous catalysts [18–20]. In particular, the use of modified *clinoptilolite* as catalysts for different catalytic reactions was extensively studied [13–23]. For example, the advantage of base-exchanged natural *clinoptilolite* catalyst for the Knoevenagel reaction was reported [16]. This zeolite demonstrated to be also an effective catalyst for skeletal isomerization of n-butenes to isobutenes [14]. The original *clinoptilolite* zeolite exhibited very low activity in *o*-xylene isomerization, due to limitation by the access of the reactant inside the zeolite channels [4]. However, the zeolite leaching with HCl led to effective catalyst for this reaction [4]. Similarly, the *clinoptilolite* treated with HCl solutions exhibited a good performance in the liquid phase isomerization of α-pinene [12]. In contrast to the HCl treatment, the *clinoptilolite* zeolite treated with NH_4Cl exhibited a low activity in the *o*-xylene isomerization [4]. This was linked with the collapse of part of the zeolite framework producing an increase in the secondary porosity, which enabled the *o*-xylene to reach acid sites [4].

Contrary to the investigation of *clinoptilolite* zeolite as catalyst, its use for supporting hydrotreating catalysts was scarcely studied [4, 18, 23]. In this sense, our previous study on the effect of the incorporation of metals (NiMoW) into this natural Mexican *clinoptilolite* zeolite, followed by sulfidation, demonstrated that those catalysts exhibited low HDS activity, which was explained as due to the low specific area (S_{BET}) of pristine zeolite [23]. It was concluded that the elimination of impurities from the pristine zeolite is needed to increase the zeolite specific surface area. Thus, the aim of present work was to remove those cations from the zeolite structure to enhance the support morphology and textural properties needed for easy reactant and products diffusion into zeolite inner structure. With this objective, acid zeolite leaching at different conditions was employed. The objectives of this acid leaching were various: (i) an increase of the zeolite specific area by chemical leaching of polyvalent cations; (ii) an increase of the catalyst stability by an increase in Si/Al ratio (due to dealumination); (iii) an increase of the catalyst acidity; (iv) the modification of the pore system by incorporation of the metal oxides into internal zeolite porous structure. The effects of zeolite acid leaching and Mo loading on the catalyst behavior of sulfided NiMo/*Clinoptilolite* catalysts were evaluated in the selective hydrodesulfurization (HDS) of 3-methylthiophene (3-MeT) reaction.

2. Experimental

2.1 Modification of the original *Clinoptilolite* by acid treatment

The natural zeolite used for supporting NiMo catalysts was a natural *clinoptilolite* zeolite from the Cuitzeo area deposit (Michoacán, Mexico). For more details on the physicochemical features of this natural mineral, the reader is addressed to Ostrooumov et al. and Huirache-Acuña et al. [3, 23]. Prior to acid leaching, the *clinoptilolite* zeolite was crushed and sieved to obtain particle size <297 mesh. Then, the zeolite was washed with deionized water and dried at 80°C overnight. Element composition of the natural zeolite was investigated by Huirache-Acuña et al. [23].

The acid leaching with 1 M nitric acid (with a proportion of 10 mL of solution per gram of zeolite) was performed at 80°C under stirring for either 24or 48 h. Those zeolite materials will be denoted hereafter as Z-1 and Z-2, respectively. After filtering, the solid was washed repeatedly with excess of distilled water until all traces of nitric acid were removed. Then, the solids were dried at 110°C for 14 h and calcined at static air conditions at 500°C for 5 h.

2.2 Preparation of bimetallic $NiMo_{(x)}$/*Clinoptilolite* catalysts

All oxide catalyst precursors were prepared by co-impregnation taking into account the isoelectric points (PIZ) of Z-1 and Z-2 supports. The measurements of isoelectric points of the *clinoptilolite* zeolite confirmed that zeolite leaching with HNO_3 during different times (24 and 48 h) led to very small change of the isoelectric point (**Figure 1**). An aqueous solutions of ammonium molybdate tetrahydrate ($H_{24}Mo_7N_6O_{24}.4H_2O$) and nickel acetate tetrahydrate ($C_4H_6NiO_4{\cdot}H_2O$) were used as molybdenum and nickel precursors, respectively. The corresponding amount of the metal salts of Mo and Ni were dissolved separately at room temperature in 5 mL of deionized water taking into account the isoelectric point of each zeolite. After impregnation, the catalysts were dried overnight in air at 110°C and then calcined at 500°C for 3 h. The catalysts were prepared with similar nickel loading and different Mo contents (**Table 1**). The samples with low, medium, and high molybdenum loadings are denoted hereafter as $NiMo_{(L)}$/Z-1(2), $NiMo_{(M)}$/Z-1(2), and $NiMo_{(H)}$/Z-1(2), respectively.

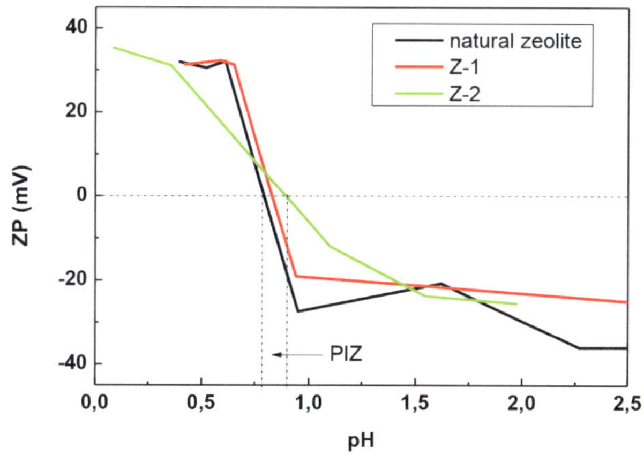

Figure 1.
Isoelectric points (PIZ) of the clinoptilolite *zeolite modified by leaching with HNO_3 during different times (24 and 48 h).*

Sample	Ni (wt.%)	Mo (wt.%)	Support treatment	Total acidity[a] (μmol NH_3/g)
Ni-$Mo_{(L)}$/Z-1	1.19	8.16	HNO_3/24 h	0.18 (0.14)
Ni-$Mo_{(L)}$/Z-2			HNO_3/48 h	0.23 (0.11)
Ni-$Mo_{(M)}$/Z-1	1.64	11.3	HNO_3/24 h	0.18 (0.14)
Ni-$Mo_{(M)}$/Z-2			HNO_3/48 h	0.27 (0.11)
Ni-$Mo_{(H)}$/Z-1	2.82	19.43	HNO_3/24 h	0.10 (0.14)
Ni-$Mo_{(H)}$/Z-2			HNO_3/48 h	0.12 (0.11)

[a]*Acidity data of the substrates are given in parenthesis.*

Table 1.
Catalyst codes, nominal metal loadings, and total acidity (from TPD-NH_3) of the calcined Ni-Mo(x)/ Clinoptilolite *samples.*

2.3 Characterization techniques

The isoelectric points (PIZ) of the *clinoptilolite* zeolite modified with different treatments were measured on a Zeta-Meter System 3.0 containing a cell electrophoretic acrylic type II-grape and platinum electrodes. Surface morphology analysis of the pure zeolites and oxide catalysts precursors was studied by SEM technique using a JEOL JSM-6060 LV microscope. The accelerating voltage employed was 15 kV. Elemental analysis was performed by Energy Dispersive X-ray Spectroscopy (EDS) Oxford Inca X-Sight coupled to a MT 1000, Hitachi apparatus. The textural properties of the oxide precursors and bare supports were determined from the adsorption-desorption isotherms of nitrogen at −196°C, recorded with an ASAP 2420 Micromeritics V2.09 gas sorption Analyzer. Prior to the experiments, the samples were degassed under a flow of argon at 350°C for 16 h. The acidity of the oxide precursors was determined by temperature-programmed desorption (TPD) of ammonia measurements conducted on a Micromeritics 2900 equipment provided with a TCD and interfaced to a data station. The morphology of fresh sulfided catalysts was investigated by high resolution transmission electron microscopy (HRTEM) using a JEOL JEM 2000FX microscope. XPS spectra of the sulfided (10% H_2S in H_2, 673 K) catalyst samples were measured at room temperature using a VG Scientific LTD system equipped with a hemispherical electron analyzer and a Mg Kα (hν = 1253.6 eV) X-ray source and ultrahigh vacuum chamber. Details of the XPS measurements by this spectrometer are reported by Huirache-Acuña et al. [23].

2.4 Catalytic activity measurements

The catalyst activity was evaluated in the reaction of hydrodesulfurization of 3-methyl-thiophene (3MeT) carried out in a micro-flow reactor at 280°C upon atmospheric hydrogen pressure. The reactor was loaded with 100 mg of catalyst (particle size between the 80 and 120 mesh) diluted with 1 g of SiC. Before reaction, the catalyst was pre-sulfided at 400°C for 1 h using a gas mixture of 15% H_2S/H_2 (flow rate of 40 mL/min). After catalyst activation, the sample was cooled down and stabilized at reaction temperature. Then, the saturation of hydrogen with 3MeT was obtained by bubbling hydrogen (70 mL min^{-1}) through a saturator containing 3MeT liquid at 20°C. The flow of 3-MeT through reactor was 1.134×10^{-6} moles × s^{-1}. The products obtained at steady state conditions were analyzed on line by GC. For each catalyst studied, steady state conditions were reached after 1 h of time on-stream reaction. All reaction products were analyzed online with gas chromatograph Agilent-7820, FID equipped with an Agilent 30 m HP-5 capillary column. The catalytic activity was expressed as total 3-MeT conversion obtained at steady-state conditions.

3. Results

3.1 Physicochemical characterization of oxide precursors

The surface morphology of the pure Z-1 and Z-2 zeolites and their respective oxide catalyst precursors were investigated by scanning electron microscopy (SEM). The SEM micrographs in **Figure 2** illustrate the influence of the different zeolite pretreatment conditions and Mo loading on the particle size and morphology of the prepared catalysts. In general, irrespectively of the time of HNO_3 treatment (24 or 48 h), both pure Z-1 and Z-2 substrates exhibit irregular and compact structure with white points resulting from the different cations impurities. After Ni and Mo oxide loading, all samples exhibit non-uniform distribution of Ni and Mo on the catalyst surface. The $NiMo_{(H)}$/Z-1 appeared to have a much better distribution of the metal oxides on the support surface than its Z-2-supported counterpart. Besides its highest metal oxide loadings, the surface of both $NiMo_{(H)}$/Z-1 and $NiMo_{(H)}$/Z-2 catalysts were not completely covered with Ni and Mo oxides. This was confirmed by the energy-dispersive X-ray spectroscopy (EDS) analysis on the different points of the samples. SEM/EDS have shown that $NiMo_{(H)}$/Z-1 and $NiMo_{(L)}$/Z-2 samples are characterized by the high Al/Si ratio (5.5 and 4.7, respectively), which might indicate that mean crystalline phase is *clinoptilolite* [3].

The textural properties of pure Z-1 and Z-2 substrates and both oxide catalysts with highest Mo loading was studied by the N_2 physisorption at −196°C. The N_2 adsorption-desorption isotherms of both parent zeolites were similar before and after metal loadings. **Figure 3(A)** shows the N_2 isotherms of both catalysts. According to IUPAC classification, all samples exhibit a combination of a type I and IV isotherms which are typical for hierarchical materials having both micro- and meso-porous structure [24]. Indeed, the H4-type of hysteresis loop is generally observed with complex materials containing both micropores and mesopores [24]. The adsorption branch of N_2 isotherm, which does not show any limiting adsorption at high P/P_0, is typical for natural zeolites having both micro- and meso-pores [3]. However, in case of natural zeolites, an increase of the relative pressure might favor also multilayer adsorption of nitrogen on the surface of impurities [3].

The natural acid-modified *clinoptilolite* zeolite contains micropores, as deduced from the shape of the N_2 sorption isotherms (**Table 2**). However, the relatively low volume of micropores (in range 0.002–0.011 $cm^3 \cdot g^{-1}$) point out that their large amount cannot be detected because the micropores are occupied by exchangeable cations [2].

As expected, an increase of leaching time of pristine zeolite from 24 to 48 h led to an increase of specific surface area (S_{BET}) from 44 to 55 $m^2 \cdot g^{-1}$ and the total pore volume from 0.066 to 0.071 $cm^3 \cdot g^{-1}$ (**Table 2**). Simultaneously, the two-fold increase of volume of micropores occurs (from 0.006 to 0.011 $cm^3 \cdot g^{-1}$) suggesting that the nitrogen access was enhanced by elimination of blocking of the pores by impurities. However, the BET specific area and total pore volume decreased again after metal loading onto surface of zeolite. In addition, the micropore volume decreases after metal oxide incorporation into both Z-1 and Z-2 supports suggesting the modification of pore opening by deposition of the large amount of the Mo species on the carrier surface (**Table 2**).

Figure 3(B) shows the pore size distributions (PSD) evaluated from the adsorption branch of nitrogen isotherm by using BJH method. For all catalysts, pore size distribution shows two distinct peaks: the first one, located about 3.6 nm, is due to the *tensile strength effect* (TSE) [25]. The second broad peak centered at 9.5 and 20.9 nm for $NiMo_{(H)}$/Z-2 and $NiMo_{(H)}$/Z-1, respectively, represents the mesopores,

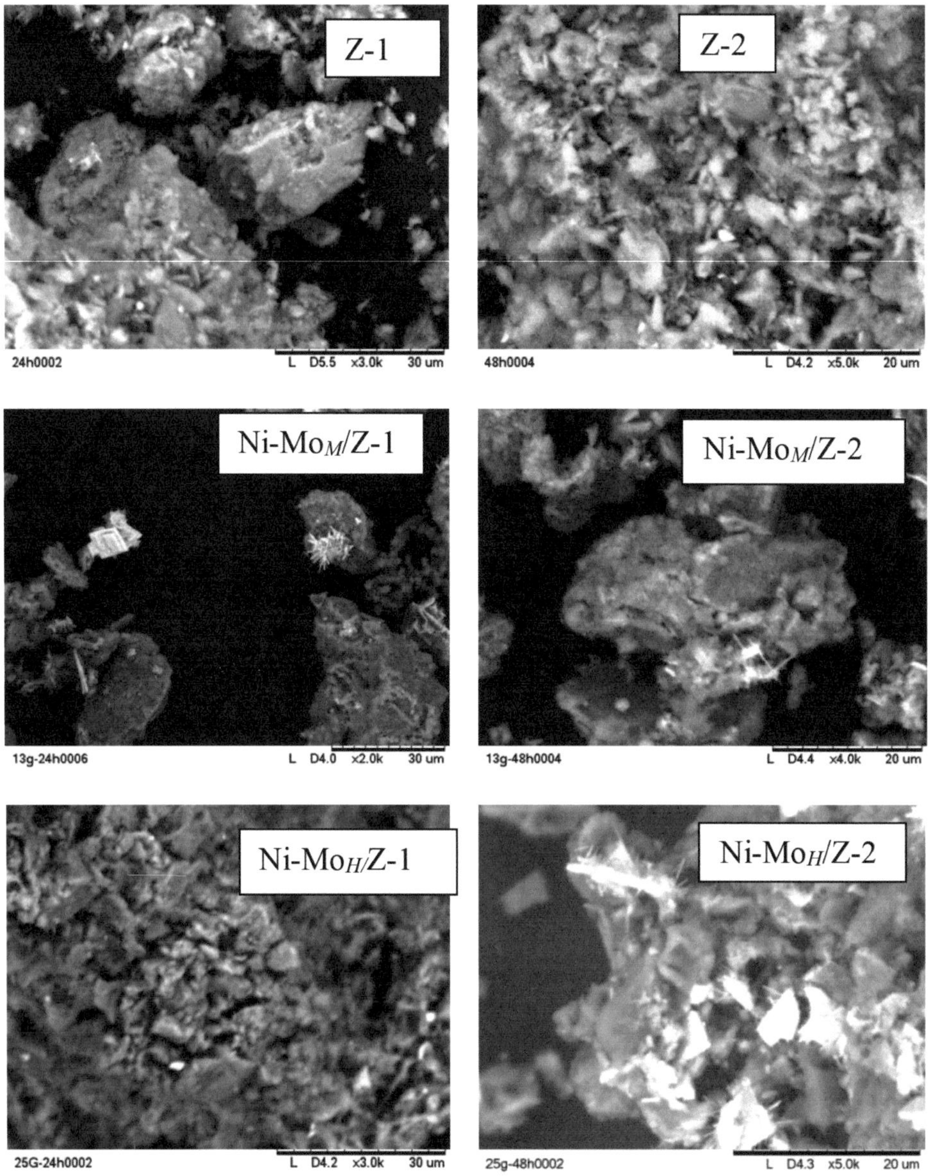

Figure 2.
SEM images of pure Z-1 and Z-2 zeolites and selected oxide catalysts' precursors.

probably intra particle porosity [26]. Noticeably, the contribution of the mesoporosity to overall catalyst porous structure is small being all catalysts mainly microporous, as deduced from the comparison of the V_{total} and V_{micro} values in **Table 2**. A comparison of the loss of S_{BET} and normalized S_{BET} values of the calcined $NiMo_{(H)}$/Z-1 catalyst with that of $NiMo_{(H)}$/Z-2 strongly suggest that the former catalyst exhibits a lower pore occlusion by guest molecules than the latter.

The acidity of the calcined catalysts was studied by temperature-programmed desorption of ammonia (TPD-NH_3). The TPD-NH_3 profiles of the most active catalysts are shown in **Figure 4**. Depending on the ammonia desorption temperature, the TPD-NH_3 profiles were mathematically fitted (Gaussian function) assuming temperature desorption ranges of 100–250°C, 250–400°C and 400–500°C, respectively, as indicative of weak, medium, and strong strength acid sites [27]. The

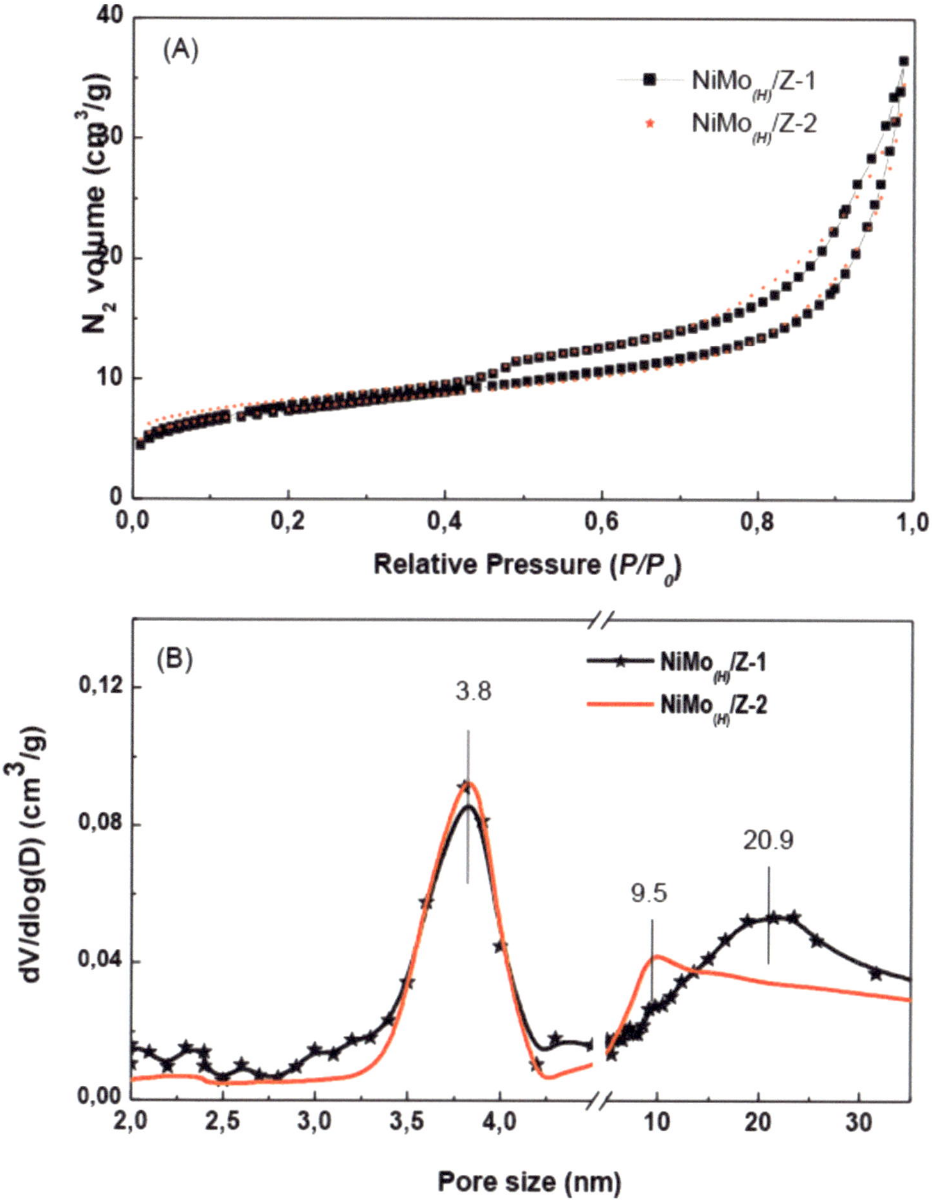

Figure 3.
N2 adsorption-desorption isotherms of the calcined Ni-Mo$_{(H)}$/Z-1(Z-2) catalysts (A) and their BJH adsorption pore size distribution (B).

weak and moderate acidity are probably linked with Brønsted acid sites, whereas the strong acid sites are probably originated by Lewis acid sites [28, 29]. As seen in **Figure 4**, the Ni-Mo$_{(L)}$/Z-2 and NiMo$_{(M)}$/Z-2 catalysts exhibited a larger amount of strong strength acid sites than their Z-1-supported counterparts suggesting the poorer metal oxides dispersion on the surface of those catalysts. The total concentration of acid sites (expressed as μmol$_{NH3}$ per g of substrate) is shown in **Table 1**. In general, the catalysts supported on Z-2 zeolite exhibit larger total acidity than their counterparts supported on Z-1 zeolite. The total acidity of the oxide precursors follows the trend: NiMo$_{(M)}$/Z-2 ≈ NiMo$_{(L)}$/Z-2 >> others. Irrespectively of the support, the strong acid sites disappeared after loading the largest Mo amount. According to Del Arco et al. [26], XRD measurements suggest that this is probably due to

Materials	S_{BET} ($m^2 g^{-1}$)	Loss of S_{BET}^{b} (%)	N_{SBET}	V_{total} ($cm^3 g^{-1}$)	V_{micro} ($cm^3 g^{-1}$)	d_p (nm)
Z-1	44	—	—	0.066	0.006	8.1
$NiMo_{(H)}$/Z-1	25	43	0.75	0.049	0.003	9.2
Z-2	55	—	—	0.071	0.011	7.6
$NiMo_{(H)}$/Z-2	25	54	0.58	0.046	0.004	10.1

[a]*Specific BET surface area (S_{BET}), total pore volume (V_{total}), micropore volume (V_{micro}) obtained from t-plot; BJH adsorption average pore diameter (d_P).*
[b]*Loss of S_{BET} after metal loading.*
[c]*Normalized specific BET surface area calculated using the equation: $NS_{BET} = S_{BET}$ of catalyst/[(1 − y) × S_{BET} of support] where S_{BET} is the specific BET surface area of the catalyst or support, and y is the weight fraction of the guest phases.*

Table 2.
*Textural properties[a], loss of S_{BET}[b] and normalized S_{BET}[c] of the calcined Ni-Mo(x)/*Clinoptilolite *catalysts.*

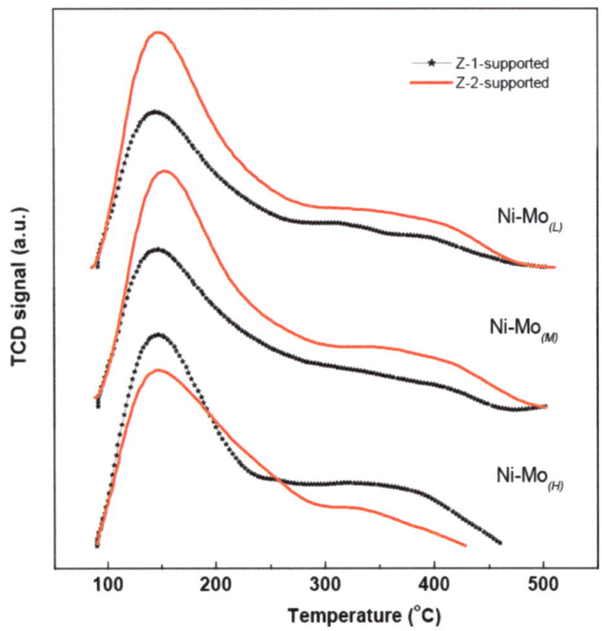

Figure 4.
*TPD-NH_3 profiles of the calcined $NiMo_{(x)}$/*Clinoptilolite *catalysts.*

formation of large MoO_3 crystallites on the support surface affecting the acidity associated with surface –OH groups.

Summarizing, taking in account all characterization data of the oxide catalyst precursors, zeolite leaching with HNO_3 during 24 h led to better support than its leaching during 48 h. This is probably because an increase of leaching time increased the support microporosity leading to a decrease of the average pore diameter (**Table 2**). After zeolite leaching and further metal loading, the structure of zeolite material was maintained, as confirmed by XRD studies by Khoshbin et al. [5]. SEM mapping analysis confirmed that main part of the metal ion impurities of natural zeolite had been displaced by H^+ ions after zeolite leaching with HNO_3 solution at 24 and 48 h. $NiMo_{(H)}$/Z-1 was the only catalyst having K^+ impurities on its support surface, as confirmed by SEM mapping (data not shown here) and chemical surface analysis by XPS (*vide infra*). Finally, the zeolite leaching with HNO_3 led to modification of the number of acid sites and their strength, as confirmed by TPD-NH_3 analysis, being more acidic the catalysts supported on zeolites leached during 24 h (**Table 1**).

3.2 Characterization of freshly sulfided catalysts

3.2.1 TEM characterizations

The HRTEM micrographs of both sulfided catalysts with highest Mo content are compared in **Figure 5**. Noticeably, the $NiMo_{(H)}$/Z-1 exhibit a larger density of Mo sulfide particles on the surface of Z-1 zeolite than its $NiMo_{(H)}$/Z-2 counterpart. The molybdenum sulfide particles on the surface of Ni-$Mo_{(H)}$/Z-1 catalyst are relatively large, and they are formed with slabs presenting a high degree of stacking faults. As seen, the slabs of the molybdenum sulfide are partially intercalated by other slabs. The interlayer spacing of 0.61 nm observed from lattice fringes can be ascribed to the (002) direction of the bulk MoS_2. However, SEM mapping of this sample (not shown here) indicated the presence of both Mo and S elements with concentration suggesting the formation of $MoS_{1.95}$ phase. For the $NiMo_{(H)}$/Z-1, the statistical analysis of about 250 particles of various HRTEM images reveals that this sample exhibits MoS_2/$MoS_{1.95}$ particles with an average size of 6.3 ± 3.1 nm and an average stacking number of about 7.0 (**Table 3**). Noticeably, the $NiMo_{(H)}$/Z-2 exhibit much lower stacking of the MoS_2 slabs than its $NiMo_{(H)}$/Z-1 counterpart (see **Figure 5(B)** and **Table 3**). Thus, both catalysts might exhibit different amounts of Mo atoms in the edge surface of molybdenum sulfide crystals [30]. To confirm this, the average fraction of Mo atoms in the edge surface of MoS_2/$MoS_{1.95}$ crystals was calculated (see f_{Mo} and f'_{Mo} values in **Table 3**) following the method used by Gutierrez and Klimova [31]. The comparison of the f'_{Mo} values, in which the layer attached to the support surface is not considered, clearly indicated that Z-1-supported catalyst contains a larger amount of those Mo atoms in the edge surface of the MoS_2/$MoS_{1.95}$ crystals than its Z-2-supported counterpart.

Noticeably, both Z-1 supported catalysts with high and medium Mo contents exhibit very similar f'_{Mo} values (**Table 3**) suggesting that the amount of Mo atoms in the edge surface of the active phase can be due to specific properties of the Z-1 support. Indeed, contrary to the Ni-$Mo_{(H)}$/Z-1 catalyst, the MoS_2 fringes of the Ni-$Mo_{(H)}$/Z-2 catalyst exhibit strong bending effect (**Figure 5**). Similar bending effect can be seen in **Figure 6** showing the HRTEM image of the Ni-$Mo_{(M)}$/Z-2 sulfide catalyst with a typical SAED pattern evidencing diffraction rings of the crystalline MoS_2 phase. The HRTEM image of this sample, the MoS_2 layers are disordered,

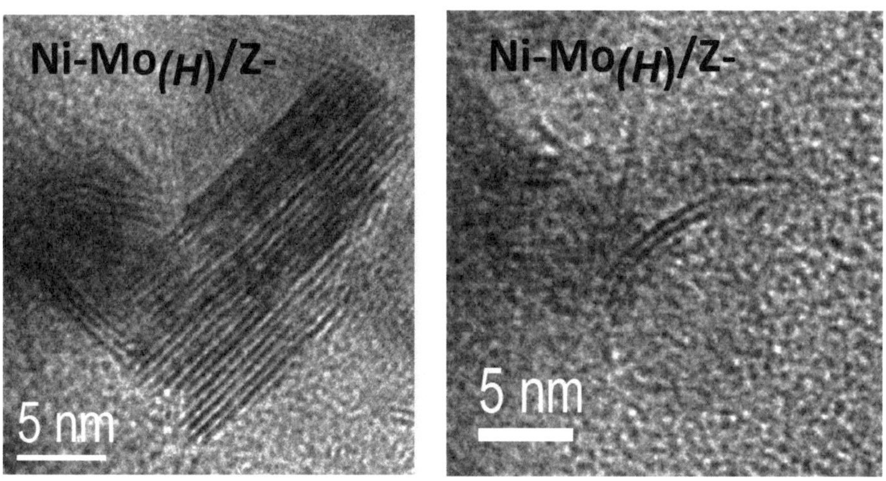

Figure 5.
HRTEM images obtained for fresh sulfided Ni-$Mo_{(H)}$/Z-1 (A) and Ni-$Mo_{(H)}$/Z-2 (B) catalysts.

Sample	*L* (nm)	*N*	f_{Mo}	f'_{Mo}
$NiMo_{(H)}$/Z-1	6.3 ± 3.1	7.0 ± 5.4	0.19	0.17
$NiMo_{(H)}$/Z-2	6.0 ± 3.6	1.2 ± 1.1	0.20	0.03
$NiMo_{(M)}$/Z-1	6.9 ± 3.8	8.4 ± 7.9	0.18	0.16

[a]*MoS_2 particle size: length (L) and number of slabs (N); f_{Mo} is average fraction of Mo atoms in the edge surface of the MoS_2 crystals; f'_{Mo} is f_{Mo} value in which the layer attached to the support was not considered [31] .*

Table 3.
Some HRTEM data[a] of the selected sulfided catalysts.

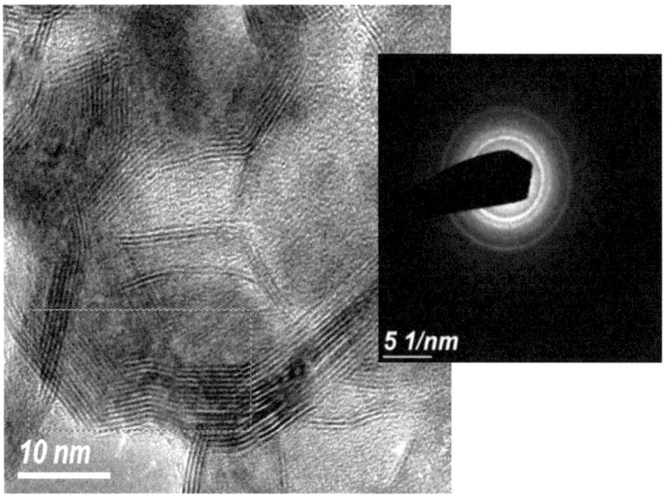

Figure 6.
High resolution TEM images of the Ni-Mo$_{(M)}$/Z-2 sulfide catalyst showing MoS2 fringes with a strong bending effect. Layers are disordered, highly stacked, and broken at the point of highest torsion. Inset: FFT analysis over the area marked by the dotted squares in panel evidencing polycrystalline structure of a large particle.

highly stacked and broken at the point of highest torsion. It is hypnotized that this can be due to the presence of impurities on the surface of Z-2 zeolite forcing the formation of the "onion-type" MoS_2 phase.

3.2.2 X-ray photoelectron spectroscopy (XPS)

Since the objective of the zeolite leaching was unblocking of its channels through dealumination and decationation, the surface exposure of undesired elements of the fresh sulfided catalysts was evaluated by XPS. For all fresh sulfided catalysts, the binding energies (BE) of same components of the zeolite carrier were: Si 2p (103.4 eV), Al 2p (74.5–75.0 eV), Fe $2p_{3/2}$ (710.6–711.1 eV), and F 1 s (685.5–685.8 eV). Interestingly, the Si 2p core level peak of all samples was close to 103.4 eV, which is characteristic of O-Si-O bonds in SiO_2. Thus, the chemical environment of silicon ions was not affected by the presence of Al^{3+}, Fe^{3+}, F^-, and K^+ ions. Considering data shown in **Table 4**, the impurities present on the support surface can be K_2O, Al_2O_3, Fe_2O_3, and SiF_4. The surface exposure of the Al^{3+} ions on the Z-1 zeolite decreases according at the trend: Ni-Mo$_{(L)}$/Z-1 > Ni-Mo$_{(M)}$/Z-1 > Ni Mo$_{(H)}$/Z-1.

The sulfiding behavior of Ni and Mo on the *clinoptilolite*-supported catalysts was investigated by in situ XPS. The effect of support was evaluated by comparison of the sulfiding behavior of $NiMo_{(H)}$ samples, whereas the effect of

Sample	Al/Si at	K/Si at	Fe/Si at	F/Si at	(Al + Fe + F)/Si
$NiMo_{(L)}$/Z-1	0.217	—	0.054	0.096	0.367
$NiMo_{(M)}$/Z-1	0.198	—	0.056	0.051	0.305
$NiMo_{(H)}$/Z-1	0.126	0.029	0.058	0.053	0.237
$NiMo_{(H)}$/Z-2	0.145	—	0.051	0.051	0.247

Table 4.
Surface atomic ratios of the undesired elements of the Clinoptilolite*-based NiMo sulfide catalysts.*

Catalysts	$Mo3d_{5/2}$	Ni $2p_{3/2}$	S 2p
$NiMo_{(L)}$/Z-1	228.9	852.4 (40) 855.8 (60)	161.9
$NiMo_{(M)}$/Z-1	228.9	853.2 (40) 856.0 (60)	161.8
$NiMo_{(H)}$/Z-1	228.8	853.2 (40) 856.0 (60)	161.6
$NiMo_{(H)}$/Z-2	229.0	852.5 (38) 855.7 (62)	161.8

[a]*In parentheses are peak percentages.*

Table 5.
Binding energies (eV) of core electrons[a] *of* Clinoptilolite*-based NiMo sulfide catalysts.*

Mo loading was evaluated by comparison of the Z-1-supported catalysts. The binding energies (eV) of the principal peaks are summarized in **Table 5** and the percentages of different species are given in parentheses. Concerning the Mo 3d core level spectra (not shown here), two doublet peaks containing the corresponding $3d_{5/2}$ and $3d_{3/2}$ components from the spin-orbit splitting suggested the presence of two types of molybdenum species. For all catalysts, the lower Mo $3d_{5/2}$ binding energy at 228.5–228.8 eV is similar to that reported for Mo_2S_3 phase (228.7 eV) with a Mo oxidation state of +3 [32]. Noticeably, the binding energy of the S $2p_{3/2}$ peak observed for the $NiMo_{(H)}$ catalysts at 161.6–161.9 eV is a little too low to be considered of sulfide (S^{2-}) ions present in the MoS_2 phase (162.4 eV) [33]. Thus, regardless of the support (Z-1 or Z-2) and Mo loading, the XPS results suggest that the sulfidation of molybdenum species located on the support surface is complete. Conversely to Mo, an important proportion of nickel species remains unsulfided, as deduced from two contributions of the Ni $2p_{3/2}$ component: one at 852.6–853.2 eV associated with nickel sulfide species and another one at 855.9–856.0 eV due to oxidized Ni^{2+} ions [34]. The comparison of the relative intensities of the Ni $2p_{3/2}$ peaks of sulfided and unsulfided Ni species in all samples suggests the similar degree of sulfidation of Ni^{2+} in all samples (40%). The $NiMo_{(H)}$/Z-2 sample is only one showing a little lower Ni species sulfidation degree (38%). Thus, the final state of sulfidation of the catalysts was independent on the support and molybdenum loading.

Concerning the effect of support, the comparison of the Mo/Si, Ni/Si, and S/Si atomic ratios of $NiMo_{(H)}$/Z-1 and $NiMo_{(H)}$/Z-2 catalysts clearly indicated that a better Mo and Ni species surface exposure was achieved using Z-2 as support (see Mo/Si and Ni/Si atomic ratios in **Table 6**). This is not surprising because Z-2 exhibits a larger specific surface area than Z-1 (55 m^2/g vs. 44 m^2/g). By comparison of surface atomic ratios of the impurities in the Z-1 and Z-2 catalysts, one can notice higher impurities of the Z-2-based catalysts.

Sample	Mo/Si at	Ni/Si at		$(Ni + Mo)_{(s)}$/Si	S/Si at	$S/(Ni + Mo)_{(S)}$	S/Mo
		Ni-O	Ni-S				
$NiMo_{(L)}$/Z-1	0.049	0.014	0.009	0.058	0.116	2.0	2.04
$NiMo_{(M)}$/Z-1	0.174	0.024	0.016	0.190	0.445	2.3	2.30
$NiMo_{(H)}$/Z-1	0.080	0.010	0.007	0.087	0.177	2.0	2.06
$NiMo_{(H)}$/Z-2	0.160	0.027	0.016	0.176	0.393	2.2	2.31

Table 6.
Surface atomic ratios of active phases for the Clinoptilolite*-based NiMo sulfide catalysts.*

In contrast to Mo species, the XPS results suggest that Ni species are partially sulfided. Considering the catalyst sulfidation degree, the comparison of the S/(Ni + Mo) ratios of both catalysts strongly suggest that, on the contrary to $NiMo_{(H)}$/Z-2 catalyst, the $NiMo_{(H)}$/Z-1 exhibits non stoichiometric molybdenum sulfide phase. Contrary to Z-1 supported catalysts, the results of the XPS analysis show that the S:Mo ratio on the samples supported on Z-2 zeolite is higher than that found for MoS_2 (**Table 6**). The excess of sulfur in those samples corresponds to elemental sulfur. This is important observation because it can explain the much better catalytic response of the $NiMo_{(H)}$/Z-1 sample with respect its Z-2 supported counterpart.

In conclusion, the interpretation of the catalyst structure by XPS results appears more straightforward than by HRTEM. The main reason for difficulty of the catalyst characterization by the latter technique is related to the fact that many impurities are still present of the catalyst surface after zeolite leaching with HNO_3 which leads to a limited number of TEM zones where molybdenum and nickel sulfides are both clearly visualized.

3.2.3 Hydrodesulfurization of gasoline model compound

The current desulfurization technology for gasoline production requires highly selective catalysts to remove S-compounds without excessive olefins hydrogenation. The latter requirement is especially important for preservation of the research octane number (RON) [1, 2]. Thus, the catalyst needed to the hydrotreating of naphtha stream needs to possess moderate hydrogenation properties. In this work, the activity and selectivity of the sulfided $NiMo_{(x)}$/*Clinoptilolite* catalysts were evaluated in the hydrodesulfurization (HDS) of 3-methylthiophene (3-MeT) as gasoline model compound. The reaction was carried out in a flow reactor under atmospheric hydrogen pressure and reaction temperature of 280°C. The 3-MeT conversion was low (0.8–5.3%) in order to avoid diffusional problems and to compare the catalysts under kinetic conditions. The intrinsic 3-MeT HDS activities at 360°C of the $NiMo_{(x)}$/Z-1 and $NiMo_{(x)}$/Z-2 catalysts (expressed as moles of 3-MeT converted per second and gram of catalyst), under steady-state conditions and reaction temperature of 280°C, are compared in **Figure 7(A)**. The specific reaction rate of the best catalyst ($NiMo_{(H)}$/Z-1) was close to that of Ni-W/SiO_2 (60×10^{-8} $mol_{3MT}\,g^{-1}\,s^{-1}$ vs. 70×10^{-8} $mol_{3MT}\,g^{-1}\,s^{-1}$) tested at the same reaction temperature (280°C) [35].

Regardless of the support and metal loading, the reaction products identified by GC were; 3-methyltetrahydrothiophene (3MTHT); 2-methylbutane-1-thiol (2MBT1); 2-methyl-1,3-butadiene (isoprene); 3-methyl-1-butene (3M1B); 2-methyl-1-butene (2M1B); and 2-methyl-2-butene (2M2B). The main products detected were 2-methyl-1-butene and isoprene. The HYD/DDS selectivities ratios of the catalysis are compared in **Figure 7(B)**. Noticeably, the hydrogenation of olefins to paraffin's did not occur because 2-methyl-1-butane (2 MB) was not produced. The

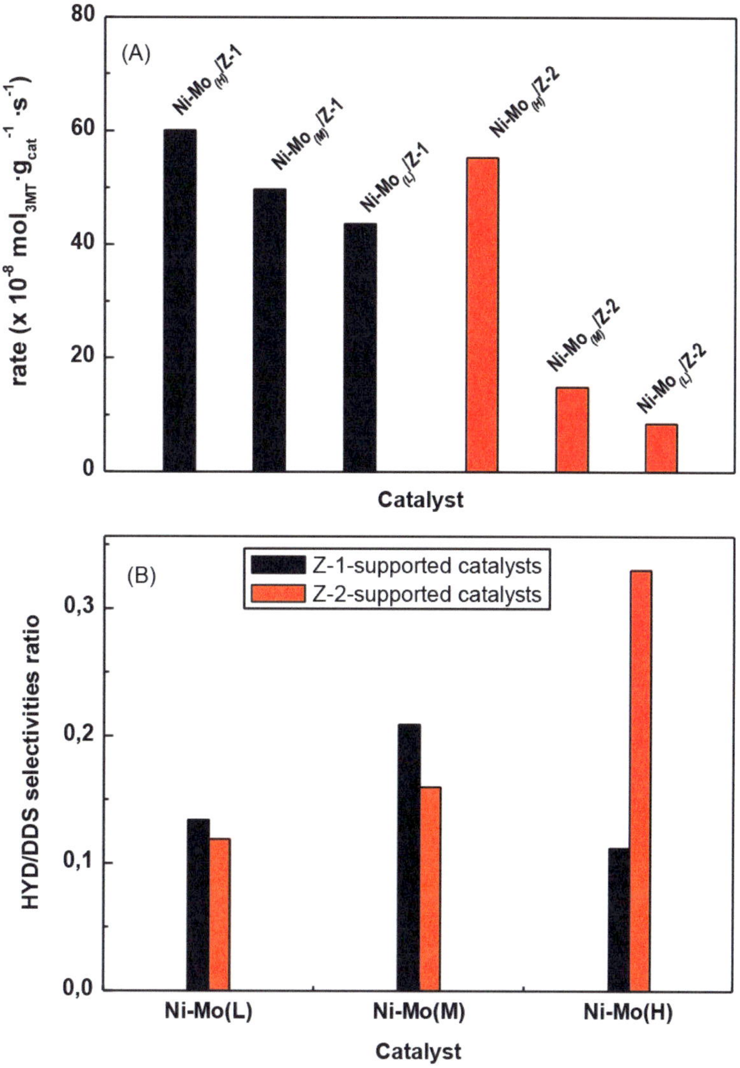

Figure 7.
*Steady state reaction rates (A) and HYD/DDS selectivities ratio (B) for HDS of 3MeT (flow reactor, T = 280°C; atmospheric H2 pressure) over sulfide NiMo$_{(x)}$/*Clinoptilolite *catalysts.*

possible reaction path network for the HDS of 3MeT over catalysts studied is shown in **Figure 8**. As seen in **Figure 8** the 3-MeT transformation occurs via hydrogenation (HYD) and direct desulfurization (DDS) reaction routes. Considering the first HYD step, the double bond of 3MeT is saturated leading to formation of 3-methyl-2,3-dihydrothiophene (3M2,3DHT). This product was not detected in this work because it was quickly transformed to 3-methyl-tetrahydrothiophene (3-MTHT) and then to a mixture of 2-methyl-butane-1-tiol (2M1BT) and 3-methyl-butane-1-tiol (3M1BT) products. On the other hand, the DDS path leads to the formation of isoprene and its subsequent transformation to mixture of olefins (3M1B, 2M1B, and 2M2B). Interestingly, the formation of 2-methylbutane via hydrogenation olefins did not occur. It is important finding of this work because such hydrogenation is not desired reaction during hydroprocessing of naphtha feedstocks.

Contrary to the NiW supported on SiO_2 and γ-Al_2O_3 [35], the formation of pentenes via isomerization of the olefin mixture compounds (3M1B, 2M1B, and 2M2B) was not observed in this work. The formation of mixed olefin compounds (3M1B,

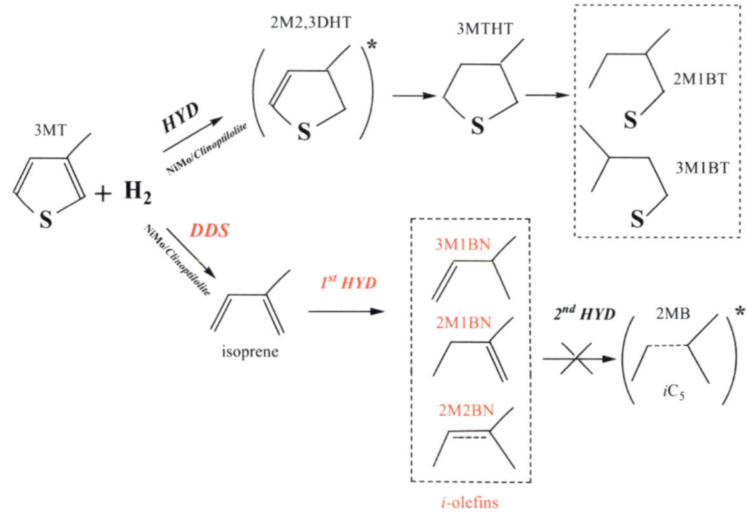

Figure 8.
*Reaction path network for the 3-methyl-thiophene (3MeT) transformation over NiMo/*Clinoptilolite *sulfide catalysts.*

2M1B, and 2M2B) was higher over the NiMo catalysts supported on Z-1 zeolite than over the counterpart Z-2 zeolite. In general, the Z-1-supported catalysts exhibit a larger formation of those mixed olefins than their counterparts supported on Z-2. Noticeably, none of the catalyst studied exhibited the formation of totally hydrogenated products. This fact is of paramount importance for production of FCC gasoline which needs the sulfur removal without excessive olefin saturation. In general, all catalysts exhibit a decrease in HYD selectivity with increasing 3-MeT conversion, as it was observed previously for sulfided Ni-Mo catalysts supported on silica-alumina [36]. This means that in the absence of thermodynamic effects, the hydrogenolysis of 3-MeT is more favorable than hydrogenation of aromatic ring. By comparing the activity and HYD/DDS selectivities ratio of the $NiMo_{(H)}$/Z-1 and $NiMo_{(L)}$/Z-1 samples in reaction at 280°C, it is clear that the increase of Mo loading led to increase of 3-MeT conversion but without changing the HYD/DDS selectivity ratio (**Figure 7(B)**). Interestingly, the largest HYD/DDS selectivities ratio is found by catalysts supported on zeolite *Clinoptilolite* leached during 48 h than their counterparts leached at 24 h.

4. Catalyst activity structure-correlation

The above activity results demonstrated clearly that the best catalyst was that prepared with largest Mo loading and supported on Z-1 carrier ($NiMo_{(H)}$/Z-1). This raise the question of why this catalyst shows higher HDS activity than its counterpart $NiMo_{(H)}$/Z-2 despite their similar nickel and molybdenum contents. To answer this question, the oxide precursors and fresh sulfided catalysts were characterized by different techniques to investigate their morphology, textural properties, acidity, and sulfidability.

In this work, the necessity of natural zeolite modification by leaching was clearly demonstrated. The objective of such leaching was to open porous structure of the natural zeolite by elimination of undesirable cations occluding the pores. For all catalysts studied, the zeolite leaching with HNO_3 during 24 h led to much better support material than its leaching during 48 h. In this sense, the pore size distribution (from N_2 physisorption measurements) confirmed that an increase of leaching time from 24 to 48 h led to an increase of the support mesoporosity, clearly observed in the 5–35 nm region of the pore size distribution (**Figure 2(A)**). Both Ni and Mo sulfide phases are poorly dispersed and distributed mainly on the external zeolite surface (from XPS), although small amounts of molybdenum sulfides can be also located within the internal support porous structure. The selected zeolite leaching conditions were not so strong in order to avoid destruction of the zeolite structure. Unfortunately, the XPS characterization confirmed that using those mild leaching conditions (HNO_3, 80°C, 24/48 h), it was impossible to eliminate totally Al^{3+}, K^+, Fe^{3+}, and F^- ions from the internal porous structure of the *Clinoptilolite* zeolite. In this sense, it was observed that not all ions have inhibition effect on the catalyst activity.

The higher activity shown by Ni-$Mo_{(H)}$/Z-1 can be attributed to the lowest amount of extraframework (EXFAL) Al atoms occluded in the internal cavities of, as shown by lowest Al/Si atomic ratio among the catalysts studied (**Table 4**). Interestingly, the 3-MeT conversion at 280°C decreased linearly with an increase of Al/Si atomic ratio of Z-1-supported samples. This is probably because the occlusion of Al atoms in the cavities of zeolite increased the catalyst acidic properties favoring the catalyst deactivation induced by acid sites. The best catalytic response of the $NiMo_{(H)}$/Z-1 catalyst can be linked also to the presence of K^+ cations decorating support surface, as this was unique sample among all catalysts studied having K^+ ions on the support surface (**Table 4**). Similar observation was reported previously for the CoMo HDS catalysts supported on aluminosilicate doped with Na [37]. In addition, the presence of potassium species on the catalyst surface might increase the dispersion of Mo phase as well as to influence the coordination state of Mo^{6+} and Ni^{2+} ions in this sample, as it was observed by Raman spectroscopic study [38]. In this sense, it was found that the Ni-Mo/γ-Al_2O_3 sample doped with K^+ exhibited a lower amount of polymeric Mo-O species because potassium provoked their partial transformation into monomeric $Mo^{6+}{}_{(Th)}$ ones. The formation of new complex K-Ni-Mo-O species was proposed [38]. In line with this, it has been found that, contrary to tetrahedral nickel species, the formation of octahedral nickel species in the oxide precursor was beneficial for catalyst activity. Finally, the SEM characterization of the fresh sulfided $NiMo_{(H)}$/Z-1 catalyst suggested the formation of the nonstoichiometric $MoS_{1.95}$ phase, which was postulated as be far more active than the stoichiometric MoS_2 [39]. This is because of anionic vacancies in the molybdenum sulfide lattice of the latter compound with respect to the stoichiometric one. Thus, the deliberate synthesis of a nonstoichiometric MoS_2 is desirable. In addition, the $NiMo_{(H)}$Z-1 catalyst contain a larger amount of Mo atoms in the edge surface of the MoS_2/$MoS_{1.95}$ crystals than its Z-2-supported counterpart, as deduced from HRTEM (**Table 3**). As a consequence, the DDS route of 3MeT transformation have been favored without excessive olefins hydrogenation, as deduced from its lowest HYD/DDS ratio among the catalysts studied (**Figure 7(B)**).

5. Conclusions

The HDS of 3-methylthiophene (3MeT) over NiMo sulfide catalysts supported on natural Mexican zeolite (*Clinoptilolite*) modified by treatment with HNO_3 during 24 and 48 h was performed in a flow reactor under atmospheric hydrogen pressure. The best catalyst was the one loaded with a largest Mo wt.% and supported on HNO_3-treated zeolite during 24 h. The HDS activities of the catalysts were clearly influenced by the forms of MoS_2 slabs, being formation of "onion-type" MoS_2 negative for the catalyst activity. Respect to selectivity, the results indicated that catalysts prepared on Z-1 zeolite had lowest hydrogenation ability, in comparison with the counterpart catalysts supported on Z-2 zeolite. The higher olefin formation on catalysts supported on Z-1 zeolite could be interesting for the tailoring novel catalysts since they could be catalysts for the production of gasolines with a greater octane number.

Acknowledgements

Drs. R. Guil-López and B. Pawelec acknowledges the financial support of the Spanish Ministry of Science, Innovation and Universities (Project CTQ2016-76505-C3-1). Dr. R. Huirache-Acuña acknowledges the financial support of CIC-UMSNH 2019-2020 Project.

Conflict of interest

The authors declare no conflict of interest.

Nomenclature

HYD	hydrogenation reaction route
DDS	direct desulfurization reaction route
2 M-3,3-DHT	2-methyl-2,3-dihydrothiophene
3MTHT	3-methyltetrahydrothiophene
2M1BT	2 methyl 1-butanethiol
3M1BT	3-methyl 1-butanethiol
Isoprene	2-methyl-1,3-butadiene
3M1BN	3-methyl 1-butene
2M1BN	2-methyl 1-butene
2M2BN	2-methyl-2-butene
2MB	2-methylbutane

Author details

Cristina Farías Rosales[1], Rut Guil-López[2], Marisol Faraldos[2], Rafael Maya Yescas[1], Trino Armando Zepeda[3], Barbara Pawelec[2]* and Rafael Huirache-Acuña[1]*

1 Facultad de Ingeniería Química, Universidad Michoacana de San Nicolás de Hidalgo, Ciudad Universitaria, Morelia, Michoacán, México

2 Institute of Catalysis and Petroleum Chemistry, CSIC, Madrid, Spain

3 Centro de Nanociencias y Nanotecnología – UNAM, Ensenada, B.C., México

*Address all correspondence to: bgarcia@icp.csic.es and rafael_huirache@yahoo.it

References

[1] Hossein F, Majid V, Touray T. Desulfurization of gas oil by modified clinoptilolite. Iranian Journal of Chemistry and Chemical Engineering. 2007;**26**(2):19-25. DOI: 1021-9986/07/2/19

[2] Korkuna O, Leboda R, Skubiszewska-Zieba J, Vrublevska T, Grunko VM, Ryczkowski J. Structural and physicochemical properties of natural zeolites: Clinoptilolite and mordenite. Microporous and Mesoporous Materials. 2006;**87**: 243-254

[3] Ostrooumov M, Cappelletti P, de' Gennaro R. Mineralogical study of zeolite from new Mexican deposits (Cuitzeo area, Michoacan, Mexico). Applied Clay Science. 2012;**55**:27-35. DOI: 10.1016/j.clay.2011.09.011

[4] Arcoya A, González JA, Travieso N, Seoane XL. Physicochemical and catalytic properties of a modified natural clinoptilolite. Clay Materials. 1994;**29**:123-131

[5] Khoshbin R, Haghighi M, Asgari N. Direct synthesis of dimethylether on the admixed nanocatalysts of CuO-ZnO-Al_2O_3 and HNO_3-modified clinoptilolite at high pressures: Surface properties and catalytic performance. Materials Research Bulletin. 2013;**48**:767-777. DOI: 10.1016/j.materresbull.2012.11.057

[6] Chmielewska E, Pilchowski K. Surface modifications of natural clinoptilolite-dominated zeolite for phenolic pollutant mitigation. Chemical Papers. 2006;**60**:98-101. DOI: 10.2478/s11696-006-0018-8

[7] Barrer RM, Makki MB. Molecular sieve sorbents from clinoptilolite. Canadian Journal of Chemistry. 1964;**42**:1481-1487. DOI: 10.1139/v64-223

[8] Tihmillioglu F, Ulku S. Use of clinoptilolite in ethanol dehydration. Separation Science and Technology. 1996;**31**:2855-2865. DOI: 10.1080/01496399608000832

[9] Abdulkareem SA, Muzenda E, Afolabi AS, Kabuba J. Treatment of clinoptilolite as an adsorbent for the removal of copper ion from synthetic wastewater solution. Arabian Journal for Science and Engineering. 2013;**38**:2263-2272. DOI: 10.1007/s13369-012-0505-x

[10] Yasyerli S, Ar I, Dogu G, Dogu T. Removal of hydrogen sulfide by clinoptilolite in a fixed bed adsorber. Chemical Engineering and Processing. 2002;**41**:785-792. DOI: 10.1016/S0255-2701(02)00009-0

[11] Hutchings GJ, Themistocleous T, Copperthwaite RG. Conversion of propane and butane. US patent 5162598. November 10, 1992

[12] Dziedzicka A, Sulikowski B, Ruggiero-Mikołajczyk M. Catalytic and physicochemical properties of modified natural clinoptilolite. Catalysis Today. 2016;**259**:50-58. DOI: 10.1016/j.cattod.2015.04.039

[13] Allahverdiev AI, Irandoust S, Murzin DY. Isomerization of of α-pinene over Clinoptilolite. Journal of Catalysis. 1999;**185**:352-362. DOI: 10.1006/jcat.1999.2474

[14] Woo HC, Lee KH, Lee JS. Catalytic skeletal isomerization of n-butenes to isobutene over natural clinoptilolite zeolite. Applied Catalysis A: General. 1996;**134**:147-158. DOI: 10.1016/0926-860X(95)00216-2

[15] Lee HC, Woo HC, Ryoo R, Lee KH, Lee JS. Skeletal isomerization of n-butenes to isobutene over acid-treated natural clinoptilolite zeolites. Applied Catalysis A: General. 2000;**196**:135-142. DOI: 10.1016/S0926-860X(99)00458-5

[16] Linares CF, Goldwasser MR, Machado FJ, Rivera A, Rodríguez-Fuentes G, Barrault J. Advantages of base exchanged natural clinoptilolite as a catalyst for the Knoevenagel reaction. Microporous and Mesoporous Materials. 2000;**41**:69-77. DOI: 10.1016/S1387-1811(00)00274-2

[17] Moreno-Tost R, Santamaría-González J, Rodríguez-Castellón E, Jiménez LA, Autié MA, González E, et al. Selective catalytic reduction of nitric oxide by ammonia over Cu-exchanged Cuban natural zeolites. Applied Catalysis B: Environmental. 2004;**50**:279-288. DOI: 10.1016/j.apcatb.2004.01.019

[18] Arcoya A, Seoane XL, Soria J. Effect of iron on the deactivation of Ni/clinoptilolite catalysts by thiophene poisoning. Journal of Chemical Technology and Biotechnology. 1997;**68**:171-176. DOI: 10.1002/(SICI)1097-4660(199702)68:2<171:AID-JCTB621>3.0.CO;2-D

[19] Lee HC, Woo HC, Chung SH, Kim HJ, Lee KH, Lee JS. Effects of metal cation on the skeletal isomerization of 1-butene over clinoptilolite. Journal of Catalysis. 2002;**211**:216-225. DOI: 10.1006/jcat.2002.3732

[20] Royaee SJ, Falamaki C, Sohrabi M, Ashraf Talesh SS. A new Langmuir-Hinshelwood mechanism for the methanol to dimethylether dehydration reaction over clinoptilolite-zeolite catalyst. Applied Catalysis A: General. 2008;**338**:114-120. DOI: 10.1016/j.apcata.2008.01.011

[21] Nimwattanakul W, Luengnaruemitchai A, Jitkarnka S. Potential of Ni supported on clinoptilolite catalysts for carbon dioxide reforming of methane. International Journal of Hydrogen Energy. 2006;**31**:93-100. DOI: 10.1016/j.ijhydene.2005.02.005

[22] Özçelik Z, Soylu GSP, Boz I. Catalytic combustion of toluene over Mn, Fe and Co-exchanged clinoptilolite support. Chemical Engineering Journal. 2009;**155**:94-100. DOI: 10.1016/j.cej.2009.07.013

[23] Huirache-Acuña R, Rivera-Muñoz EM, Pawelec B, Ostrooumov M, Maya-Yescas R, Rico JL. The use of a natural Mexican zeolite as support of NiMoW sulfide hydrotreating catalysts. Catalysis Today. 2014;**220-222**:301-309. DOI: 10.1016/j.cattod.2013.07.019

[24] Thommes M, Kaneko K, Neimark AV, Oliver JP, Rodriguez-Reinoso F, Rouquerol J, et al. Physisorption of gases, with special reference to the evaluation of surface area and pore size distribution (IUPAC technical report). Pure and Applied Chemistry. 2015;**87**(9-10):1051-1069. DOI: 10.1515/pac-2014-1117

[25] Groen JC, Peffer LAA, Moulijn JA, Pérez-Ramírez J. Mesoporosity development in ZSM-5 zeolite upon optimized desilication conditions in alkaline medium. Colloids and Surfaces A: Physicochemical and Engineering Aspects. 2004;**241**:53-58. DOI: 10.1016/j.colsurfa.2004.04.012

[26] Kruk M, Jaroniec M, Ko CH, Ryoo R. Characterization of the porous structure of SBA-15. Chemistry of Materials. 2000;**12**:1961-1968. DOI: 10.1021/cm000164e

[27] Lewandowski M, Sarbak Z. The effect of boron addition on hydrodesulfurization and hydrodenitrogenation activity of NiMo/Al_2O_3 catalysts. Fuel. 2000;**79**:487-495. DOI: 10.1016/S0016-2361(99)00151-9

[28] Olivas A, Zepeda TA. Impact of Al and Ti ions on the dispersion and performance of supported NiMo(W)/SBA-15 catalysts in the HDS and HYD reactions. Catalysis Today.

2009;**143**:120-125. DOI: 10.1016/j.cattod.2008.09.001

[29] Del Arco M, Carrazan SRG, Martin C, Martin I, Rives V. A FTIR assessment of surface acidity and dispersion of surface species in titania and alumina-supported molybdena. Spectrochimica Acta. 1994;**50**:697-702. DOI: 0584-8539/94

[30] Zink N, Annal TH, Pansiot J, Yella A, Banhart F, Tremel W. In situ heating TEM study of onion-like WS_2 and MoS_2 nanostructures obtained via MOCVD. Chemistry of Materials. 2008;**20**(1):65-71. DOI: 10.1021/cm061867k

[31] Gutierrez OY, Klimova T. Effect of the support on the high activity of the (Ni)Mo/ZrO_2-SBA-15 catalyst in the simultaneous hydrodesulfurization of DBT and 4,6-DMDBT. Journal of Catalysis. 2011;**281**:50-62. DOI: 10.1016/j.jcat.2011.04.001

[32] Thompson RK, Hilsenbeck SJ, Paskach TJ, McCarley RE, Schrader GL. Pretreatment of new reduced ternary molybdenum sulfide catalysts. Journal of Molecular Catalysis A: Chemical. 2000;**161**(1-2):75-87. DOI: 10.1016/S1381-1169(00)00143-6

[33] Finster J. SiO_2 in 6:3 (stishovite) and 4:2 Co-ordination-characterization by core level spectroscopy (XPS/XAES). Surface and Interface Analysis. 1988;**12**:309-314. DOI: 10.1002/sia.740120507

[34] Lee DK, Lee HT, Lee IC, Park SK, Bae SY, Kim CW, et al. W-incorporated CoMo/γ-Al_2O_3 hydrodesulfurization catalyst: II characterization. Journal of Catalysis. 1996;**159**:219-229. DOI: 10.1006/jcat.1996.9999

[35] Díaz de León JN, Zavala-Sánchez LA, Suárez-Toriello VA, Alonso-Núñez G, Zepeda TA, Yocupicio RI, et al. Support effect of NiW catalysts for highly selective sulfur removal from light hydrocarbons. Applied Catalysis B: Environmental. 2017;**213**:167-176. DOI: 10.1016/j.catb.2017.05.014

[36] Pawelec B, Navarro RM, Campos-Martin JM, López-Agudo A, Vasudevan PT, Fierro JLG. Silica–alumina-supported transition metal sulphide catalysts for deep hydrodesulphurization. Catalysis Today. 2003;**86**:73-85. DOI: 10.1016/S0920-5861(03)00405-X

[37] Venezia AM, Raimondi F, La Parola V, Deganello G. Influence of sodium on the structure and HDS activity of Co-Mo catalysts supported on silica and aluminosilicate. Journal of Catalysis. 2000;**194**:393-400. DOI: 10.1006/jcat.2000.2953

[38] Nikolova D, Edreva-Kardjieva R, Giurginca RM, Meghea A, Vakros J, Voyiatzis GA, et al. The effect of potassium addition on the state of the components in the oxide precursor of the (Ni)(Mo)/γ-Al_2O_3 water-gas shift catalysts: FT-IR, diffuse reflectance and Raman spectroscopic studies. Vibrational Spectroscopy. 2007;**44**:343-350. DOI: 10.1016/j.vibspec.2007.03.002

[39] Lindner J, Sachdev A, Schwank J, Villa-Garcia M. Characterization and hydrodesulfurization activity studies of unpromoted molybdenum sulfides prepared by elemental solid state reaction. Journal of Catalysis. 1992;**137**(2):333-345. DOI: 10.1016/0021-9517(92)90161-A

Chapter 3

Functionalized MCM-48 as Carrier for In Vitro Controlled Release of an Active Biomolecule, L-Arginine

Anjali Uday Patel and Priyanka Dipakbhai Solanki

Abstract

The present chapter describes the synthesis, characterizations, and application of MCM-48 functionalized by an inorganic moiety, as a carrier. MCM-48 functionalized by 12-tungstophophoric acid (TPA) (TPA-MCM-48) and L-arginine was loaded into pure as well as functionalized MCM-48. Both the materials were characterized by various physicochemical techniques and evaluated for in vitro release of L-arginine at body temperature under different conditions. A study on release kinetics was carried out using first-order release kinetic model, while the mechanism were by Higuchi model. Further, to see the influence of TPA on release rate, release profile obtained from pure and functionalized MCM-48 was compared.

Keywords: MCM-48, 12-tungstophosphoric acid, L-arginine, in vitro release, kinetics and mechanism

1. Introduction

A semi essential amino acid, L-arginine is the main source of generation of NO via NO synthase (NOS), nitric oxide (NO), polyamines and agmatine, which also influence hormonal release and the synthesis of pyrimidine bases [1–4]. The three NOS isoforms have been found to be expressed in the kidney [4]. In the kidney endothelial NOS is important in the maintenance of glomerular filtration rate, regional vascular tone, and renal blood flow. The neuronal NOS (nNOS) is expressed primarily in the macula densa and participates in the control of glomerular hemodynamic via tubuloglomerular feedback and rennin release. The inducible (iNOS) is expressed in the kidney under pathological condition in the glomerular mesangium, infiltrating macrophages and tubules [5]. L-arginine is the substrate for arginases, a group of enzyme that are involved in tissue repair processes and that metabolize L-arginine to l-ornithine [6] as well as precursor for polyamine synthesis which is also involved in tissue repair and wound healing [7, 8]. During the time of stress, body does not provide sufficient amount of L-arginine for metabolic needs. Hence, under this condition, L-arginine supplementation has been considered as an adjunct treatment for restoring normal function [7, 9, 10]. It was also found that compared to oral administration, intravenous administration of L-arginine to the patients with coronary artery disease increases

IntechOpen

the bioavailability of vascular nitric oxide (NO) which shows the vasodilator effect. Further, it was found that in case of oral administration, the bio-availability of L-arginine decreases as it is utilized by arginase for the production of urea and ornithine and thus competes with NO synthase for substrate availability [1, 11], 40% L-arginine is degraded in the intestine by arginase [12]. The mentioned problem can be overcome if some carrier will be used. The advantage of using carrier is expected to provide sufficient amount of L-arginine for NO production as the carrier can effectively load the desired amount of L-arginine as well as release it in control manner.

Thus, even though, L-arginine is important and is used as drug under certain conditions, no work has been found on controlled release of L-arginine except one which deals with the adsorption of L-arginine on SBA-15 at different pH by Deng et al. [13].

As M41S family have properties like higher surface area, ordered porosity, higher adsorption capacity, biocompatibility and non-cytotoxicity, they have been effectively explored as drug delivery carrier [14] and a number of reports are available on unfunctionalized or functionalized MCM-41/SBA-15 [15–40]. At the same time, very few reports are available on MCM-48, another important member of the M41S family as drug delivery carrier [40–43]. Arruebo et al. have reported iron loaded MCM-48 as drug delivery carrier [40]. Yang et al. have reported functionalization of MCM-48 by hydroxyapatite and used as a carrier for captopril [41]. Gai et al. have reported in vitro release of captopril using MCM-48 functionalized by VO4:Eu3+ [42]. Aghaei et al. have reported in vitro release of ibuprofen using MCM-48 functionalized by hydroxyapatite [43]. Recently, Popat et al. have reported the release profile of Sulfasalazine from MCM-48. [44]. Choi et al. have reported the comparative study for release of hydrophilic dye (Rose bengal) and hydrophobic molecule [Camptothecin (CPT) and Rhodamine 6G (R6G)] from MCM-48 [45]. Berger et al. have reported comparative study for release of aminoglycoside from MCM-48 [39].

A literature survey shows that MCM-48 has been used as a carrier either in pure form or in functionalized form using organic moiety [41, 44, 46] and no reports are available on functionalization of MCM−48 by inorganic moiety.

In this regard, polyoxometalates (POMs), early transition metal oxygen anion clusters with metal in their higher oxidation states, are excellent candidates for the same as they have already been used in medicinal chemistry. The most investigated POMs are the Keggin type, represented by the general formula $[Xn^{n+}M_{12}O_{40}]^{(8-n)-}$, where Xn^{n+} is a central heteroatom (Si^{4+}, P^{5+}, etc.) and M is an addenda atom (W^{6+}, Mo^{6+}, V^{5+}, etc.). They have large number of water molecules and terminal oxygen atoms through which they can bind to the different functional groups of drug molecules. Literature survey shows that amongst all Keggin POMs, 12-tungstophosphoric acid (TPA) has been widely used in medicinal chemistry [47–51]. In the present chapter, first time we are describing the use of 12-tungstophosphoric acid (TPA) for functionalization of MCM-48.

The present chapter describes synthesis of MCM-48, its functionalization by TPA (TPA-MCM-48) and encapsulation of L-arginine into TPA-MCM-48 (L-arg_1/TPA-MCM-48) as well as their characterization using various physicochemical techniques. In vitro release of L-arginine was carried out at body temperature under static and dynamic condition and different pH. L-arginine release kinetics and mechanism was also investigated using first order release kinetic model and Higuchi model. Further, in order to see the influence of TPA on release rate, L-arginine was loaded on pure MCM-48 (L-arg_1/MCM-48) and its release study was carried out under same experimental condition.

2. Experimental

2.1 Materials

L-arginine (SRL), Tetraethyl Orthosilicate (TEOS), Cetyltrimethylammonum Bromide (CTAB), Ninhydrin, Sodium hydroxide (NaOH), 12-Tungstophosphoric acid (TPA) and Ethanol (absolute) were received from Merck. All the chemicals were of A.R. Grade and used without any further purification.

Simulated body fluid (SBF): NaCl (7.996 g), $NaHCO_3$ (0.350 g), KCl (0.224 g), $K_2HPO_4.3H_2O$ (0.22 g), $MgCl_2.6H_2O$ (0.305 g), 1 $molL^{-1}$ HCl (40 ml), $CaCl_2$ (0.278 g), Na_2SO_4 (0.071 g) and $NH_2C(CH_2OH)_3$ (6.057 g) were dissolved in small amount of water and then dilute to 1 L with distilled water [46, 52].

Simulated gastric fluid (SGF): 6.217 g of conc. HCl was place into 1 L volumetric flask and then diluted up to the mark using distilled water.

2.2 Synthesis of mesoporous MCM-48

The synthesis of MCM-48 was carried out as described in literature [53]. 2.4 g of CTAB was dissolved in 50 ml distilled water. To this clear solution of CTAB, 50 ml ethanol (0.87 mole) and 12.6 ml ammonia (32 wt%, 0.20 mole) were added. The mixture was then stirred for 10 min. After 10 min, 3.4 g TEOS (16 mole, 3.6 mL) was added to this homogeneous solution. The mixture was then stirred for 2 h. After stirring, the resulting white solid was filtered and wash with distilled water. The dried white material then calcined at 823 K for 6 h to remove the template. The resulting material was designated as MCM-48.

2.3 Synthesis of TPA functionalized MCM-48 (TPA-MCM-48)

MCM-48 was functionalized using TPA as functionalizing agent by incipient wet impregnation method [54]. 30% of TPA anchored to MCM-48 was synthesized. One g of MCM-48 was impregnated with an aqueous solution of TPA (0.3/30 g/mL of distilled water) and dried at 100°C for 10 h. The obtained material was designated as TPA-MCM-48.

2.4 Loading of L-arginine on TPA-MCM-48 and MCM-48

Loading of L-arginine was also carried out by incipient wet impregnation method where 1 g of TPA-MCM-48 and MCM-48 were impregnated with an aqueous solution of L-arginine (0.1 g/10 mL) and pH of the mixture was adjusted up to 4 using 0.1 M HCl solution and mixture was dried at 100°C for 5 h. The obtained material was designated as L-arg_1/TPA-MCM-48. Loading amount was also confirmed by thermal analysis which shows 0.1 g of L-arginine was loaded per g of both materials (TPA-MCM-48 and MCM-48). All the experiments were carried out three times for finding % errors with in data.

2.5 Preparation of calibration curve for determination of L-arginine

To obtained calibration curve of L-arginine, stock solution was prepared by dissolving 1 g of it in 100 ml SBF (1 mg/mL). Series of 25 mL Nessler's tubes containing 0.15–0.4 mL of this stock solution was diluted up to the 4 mL using distilled water. 1 mL of acetate buffer (pH 5.5) and 1 mL ninhydrin reagent (10% ninhydrin solution in ethanol) were added to diluted solutions. All the tubes were placed in

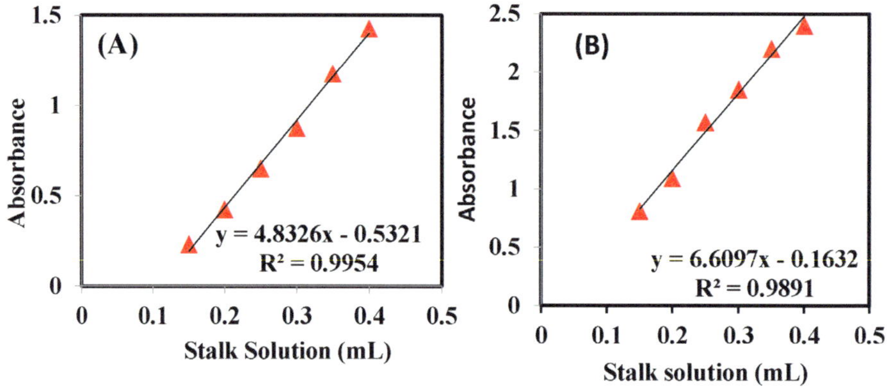

Figure 1.
Calibration curve of L-arginine in (A) SBF and (B) SGF.

boiling water bath up to 15 min for completion of reaction which was seen by the formation of purple color. The solutions were cooled under tap water, followed by addition of 1 mL of 50% ethanol and absorption was taken at 570 nm using Systronics UV–Visible-spectrophotometer. Similarly, calibration curve was obtained in SGF (**Figure 1**).

2.6 Characterization

The FTIR spectra of all materials were obtained by using KBr palate on Perkin Elmer instrument. TGA of the materials were carried out using a Mettler Toledo Star SW 7.01 instrument under nitrogen atmosphere from 30 to 570°C at the heating rate of 10°C/min. Adsorption–desorption isotherms of MCM-48, TPA-MCM-48 and L-arg$_1$/TPA-MCM-48 were recorded on a Micromeritics ASAP 2010 Surface area analyzer at liquid nitrogen temperature. From the adsorption desorption isotherms, specific surface area was calculated using BET method. The XRD pattern was obtained on PHILIPS PW-1830, with Cu Kα radiation (1.54 Å) and scanning angle from 0° to 10°. TEM analysis was carried out on JEOL (JAPAN) TEM instrument (model-JEM 100CX II) with accelerating voltage of 200 kV. The samples were dispersed in ethanol and ultrasonicated for 5–10 min. A small drop of the sample was then taken in a carbon coated copper grid and dried before viewing. ^{31}P MAS NMR spectra were recorded by BRUKER Avance DSX-300, at 121.49 MHz using a 7 mm rotor probe with 85% phosphoric acid as an external standard. The spinning rate was 5–7 kHz. Catalyst samples, after treatment were kept in a desiccator over P_2O_5 until the NMR measurement. The ^{29}Si NMR spectra were recorded at Mercury Plus 300 MHz using a 5 mm Dual Broad Band rotor probe with TMS as an external standard.

2.7 In vitro release study of L-arginine

In vitro release profile of L-arginine was obtained by soaking 0.5 g of L-arginine loaded materials in 62.5 mL of SBF (0.8 mg of L-arginine in 1 mL SBF) at temperature of 37°C with 200 rpm. At predetermined time interval, 0.5 mL of released fluid was taken and immediately equal amount of fresh SBF was added to maintain the constant volume. This release fluid was analyzed for L-arginine content by treating it with 10% ninhydrin solution at 570 nm using UV–Visible spectrophotometer (PerkinElmer). Same study was also carried under static condition as well as at different pH (pH 7.4 and pH 1.2). All the experiments were repeated three times.

3. Result and discussion

3.1 Material characterization

FTIR spectrum of MCM-48 shows a broad band around 1100 and 1165 cm^{-1} corresponding to asymmetric stretching of Si-O-Si (**Figure 2b**), the bands at 640 and 3448 cm^{-1} are corresponding to symmetric vibrations of Si-O-Si and stretching vibration Si-OH band, respectively, in good agreement with the reported one [55]. The presence of typical bands of TPA at 982 cm^{-1} corresponding to Vas vibration of $W = O_d$, and 897 cm^{-1} for stretching vibrations of W-O-W, in TPA-MCM-48 [56] suggest the primary structure of TPA is remain intact even after functionalization of MCM-48. The absence of vibration band at 1080 cm^{-1} (Vs stretching of P-O) of TPA may be due to superimposition on the bands of MCM-48. The FTIR spectrum of L-arginine shows bands around 3151, 1680, 1574 cm^{-1} corresponds to N–H stretching vibration, NH_2 in plan bending vibration and C=O stretching vibration [57]. FTIR spectrum of L-arg$_1$/TPA-MCM-48 (**Figure 2d**) shows entire bands corresponding to TPA-MCM-48 with lower intensity suggesting multiple adsorption of it into TPA-MCM-48 which is further confirmed by Nitrogen adsorption–desorption surface area analysis (**Figure 3** and **Table 1**). It shows some additional bands at 3183 cm^{-1} due to the N-H stretching, 1690 cm^{-1} due to NH_2 in plane bending vibration and 1552 cm^{-1} due to C=O stretching vibration. The observed Shifting in the bands correspond to C=O group and N-H group indicate the interaction of L-arginine to TPA-MCM-48 through these functional group.

TGA analysis of pure MCM-48, TPA-MCM-48 and L-arg$_1$/TPA-MCM-48 are shown in **Figure 3**. MCM-48 shows initial weight loss of 0.7% up to 100°C. This initial weight loss may be due to adsorption of physically adsorbed water molecules. After that, no further weight loss is observed.

TGA curve of TPA-MCM-48 shows initial weight loss of 10–13% up to 100°C which indicate presence of adsorbed water, while the Second weight loss of 1–2% between 200 and 300°C corresponds to the loss of water of crystallization of Keggin

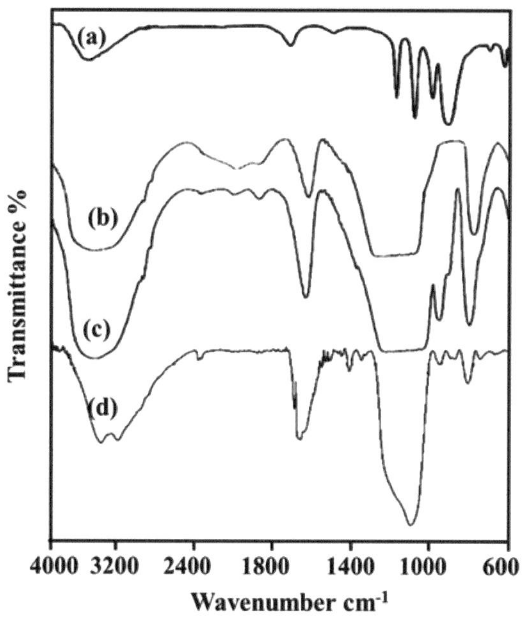

Figure 2.
FTIR spectra of (a) TPA, (b) MCM-48, (c) TPA-MCM-48 and (d) L-arg$_1$/TPA-MCM-48.

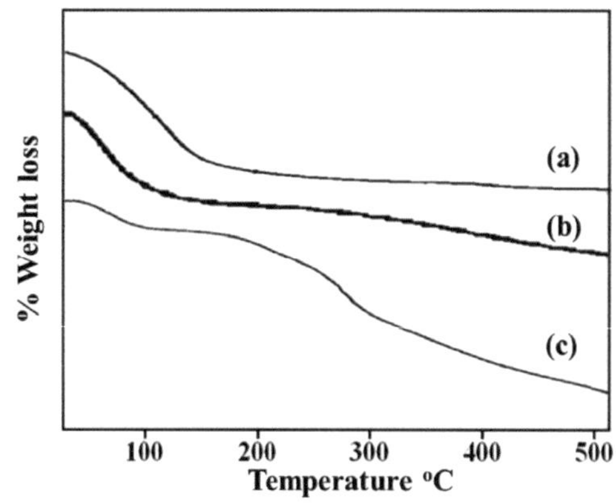

Figure 3.
TGA analysis (a) MCM-48, (b) TPA-MCM-48 and L-arg1/TPA-MCM-48.

Materials	Specific surface area (m2/g)	Pore volume (cm3/g)
MCM-48	1141	0.67
TPA-MCM-48	566	0.22 [R,S]
L-arg_1/TPA-MCM-48	68	0.10
L-arg1/MCM-48	109	0.25

Table 1.
Textural properties of MCM-48, TPA-MCM-48, L-arg1/TPA-MCM-48 and L-arg1/MCM-48.

ion. A gradual weight loss from 300 to 500°C was observed due to the difficulty in removal of water present in TPA molecules inside the channels of MCM-48. This types of inclusion can cause the stabilization of TPA molecules inside the channels of MCM-48.

TGA curve of L-arg_1/TPA-MCM-48 shows initial weight loss of 2.45% up to 100°C and further weight loss of 9.93% from 200 to 500°C. Initial weight loss may be due to the presence of adsorbed water. Weight loss from 200 to 490°C may be because of the removal of L-arginine from TPA-MCM-48.

Figure 4 shows nitrogen adsorption–desorption isotherm of MCM-48, TPA-MCM-48 and L-arg_1/TPA-MCM-48 and textural parameters are shown in **Table 1**. Isotherms of MCM-48 is type IV in nature and according to IUPAC classification it is the characteristic of mesoporous solids. This also suggests that functionalization as well as loading of L-arginine does not alter the structure of MCM-48 Decrease in surface area and pore volume is observed after functionalization of MCM-48 by TPA suggests the strong interaction of TPA with MCM-48. Further decrease in surface area and pore volume is observed in case of L-arg_1/TPA-MCM-48 which suggests the encapsulation of L-arginine into the mesoporous channel of TPA-MCM-48.

Figure 5 shows Low angle powder XRD of MCM-48, TPA-MCM-48 and L-arg_1/TPA-MCM-48. The XRD pattern of MCM-48 shows main characteristic peaks at 2.3° which indicates the presence of intact structure of MCM-48. It also shows broad shoulder at 3.0° corresponding to a plane 211 and 220, respectively. Several peaks in the range of 4–5° diffraction angles correspond to the reflections of 400, 321 and 420 planes of a typical MCM-48 meso-structure with Ia3d cubic symmetry. It is well known that the low angle XRD pattern are sensitive to pore filling and loaded materials show lowered intensity of characteristic peak compared to pure one. XRD

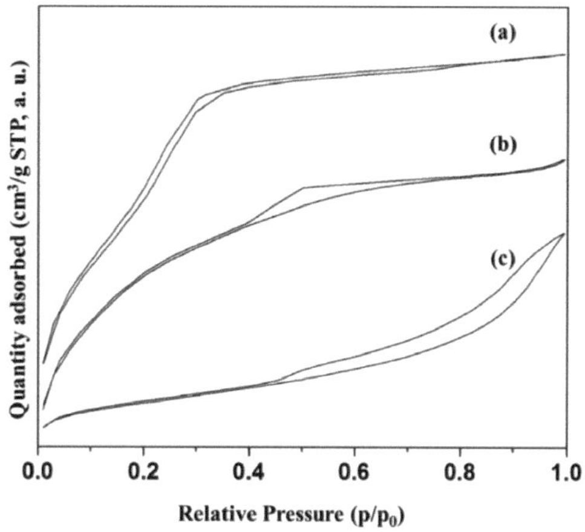

Figure 4.
Nitrogen adsorption–desorption isotherms of (a) MCM-48, (b) TPA-MCM-48 and (c) L-arg$_1$/TPA-MCM-48.

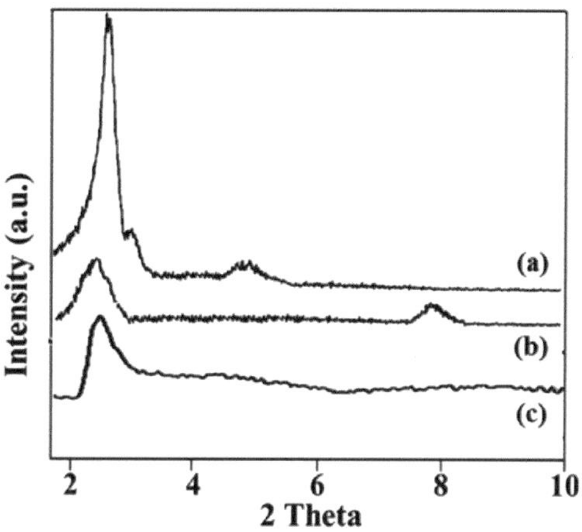

Figure 5.
Low angle powder XRD of (A) MCM-48, (B) TPA-MCM-48 and (C) L-arg$_1$/TPA-MCM-48.

pattern of TPA-MCM-48 and L-arg$_1$/TPA-MCM-48 shows characteristics peak at $2\theta = 2.3°$ with lower intensity. Absence of any other peak indicates the insertion as well as homogeneous distribution of L-arginine into the mesoporous channels of TPA-MCM-48. In addition to this, disappearance of secondary peak at $2\theta = 3–5°$ in case of L-arg$_1$/TPA-MCM-48 was observed. This is because further loading of L-arginine into functionalized MCM-48 may block the channels which have already been confirmed by BET analysis.

Figure 6 shows TEM images of MCM-48, TPA-MCM-48 and L-arg$_1$/TPA-MCM-48 at 50 nm resolution. The TEM image of MCM-48 shows very well ordered pore system with spherical morphology. It is well known in literature, that MCM-48 has a three dimensional pore system with two non-intersecting gyroidal pores. The TEM image of TPA-MCM-48 shows similar spherical morphology with porous system suggesting intact structure of MCM-48 even after functionalization. The TEM images of L-arg$_1$/TPA-MCM-48 also show spherical morphology and porous structure. Absence of any aggregation in TEM images of L-arg$_1$/TPA-MCM-48 suggests the homogeneous dispersion of L-arginine on TPA-MCM-48. Further the structure of TPA-MCM-48 remains intact even after loading of L-arginine.

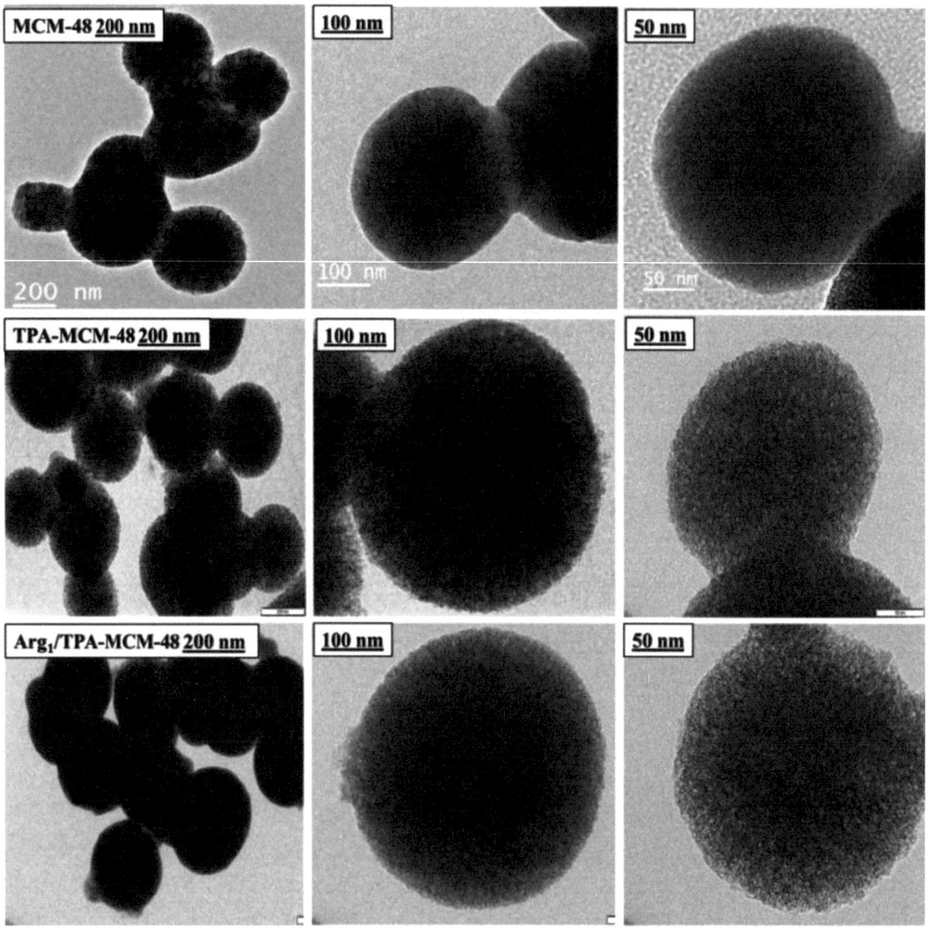

Figure 6.
TEM images of MCM-48, TPA-MCM-48 and L-arg$_1$/TPA-MCM-48 at 200, 100 and 50 nm resolution.

^{31}P MAS NMR is the most important method to study chemical environment around the phosphorous in heteropoly compounds. The ^{31}P MAS NMR spectra of TPA-MCM-48 and L-arg$_1$/TPA-MCM-48 are shown in **Figure 7**. The pure TPA shows single peak at −15.62 ppm and is in good agreement with the reported one [58]. The ^{31}P MAS NMR spectra of TPA-MCM-48 shows two peak at −12.27 and 3.060 ppm. The observed shift from −15.62 to −12.27 ppm is attributed to the strong interaction of MCM-48 with that of TPA as well as the presence of TPA inside the MCM-48. These results are in good agreement with reported one [59].

Further shift in this peak is observed from −12.27 to −9.24 ppm for L-arg$_1$/TPA-MCM-48 which also suggest the interaction of TPA with that of L-arginine which was further confirmed by ^{29}Si MAS NMR.

^{29}Si MAS NMR is a useful technique to study the chemical environment around the silicon nuclei in the mesoporous materials. **Figure 8** represents the ^{29}Si MAS NMR spectra of MCM-48, TPA-MCM-48 and L-arg$_1$/TPA-MCM-48. A broad peak of MCM-48 between −90 to −125 ppm observed which can be attributed to three main part of the peak with chemical shift at −92, −99 and − 108 ppm (**Figure 8a**, **Table 2**). These signals resulted from Q2 (−92 ppm), Q3 (−99 ppm) and Q4 (−108 ppm) silicon nuclei.

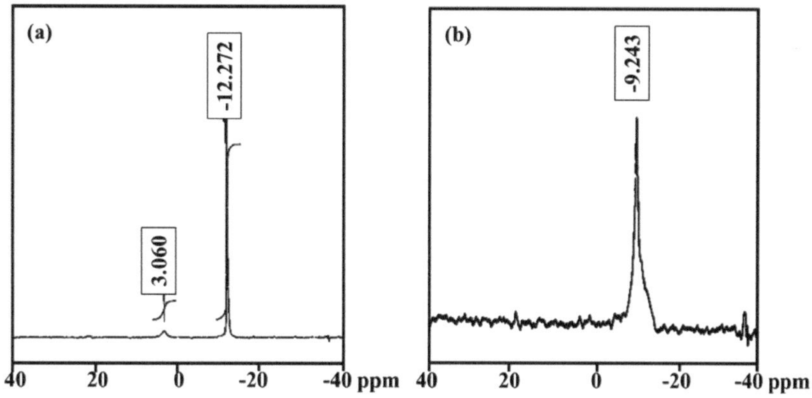

Figure 7.
^{31}P MAS NMR of (a) TPA-MCM-48 and (b) L-arg$_1$/TPA-MCM-48.

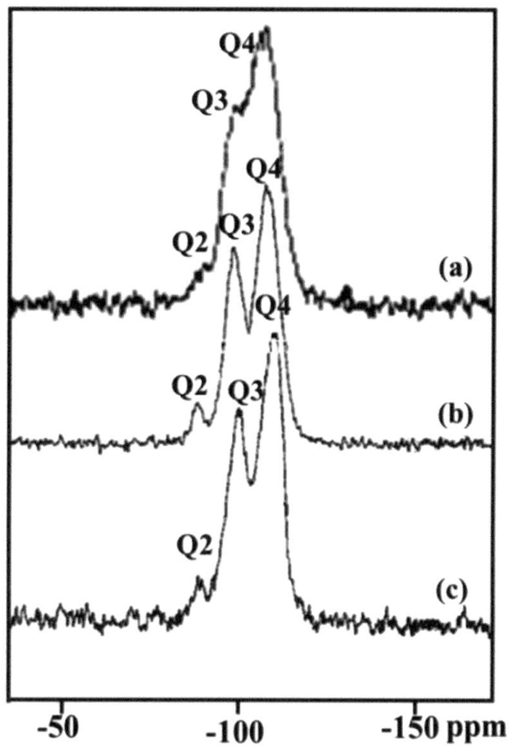

Figure 8.
(a) MCM-48, (b) TPA/MCM-48 and (c) L-arg$_1$/TPA-MCM-48.

All the three, Q2, Q3 and Q4 bands are observed in NMR spectra of TPA-MCM-48 (**Figure 8b**) suggesting the intact structure of MCM-48 even after functionalization. However, significant shift is observed in Q2 band from −92 to −88 (**Table 2**) suggests the interaction of Si-OH group of MCM-48 with TPA.

The ^{29}Si MAS NMR spectra of L-arg$_1$/TPA-MCM-48 show characteristic peaks (**Figure 8c**) of TPA-MCM-48 (**Table 2**). No significant shift in the Q2, Q3 and Q4 bands suggests that the structure of TPA-MCM-48 remains intact even after loading of L-arginine. This further confirms that the L-arginine molecules selectively bind with TPA and not to the surface Si-OH group of MCM-48.

Sr. No.	Materials	29Si MAS NMR data		
		Q2 ppm	Q3 ppm	Q4 ppm
1.	MCM-48	−92	−99	−108
2.	TPA-MCM-48	−88	−98	−108
3.	L-arg_1/TPA-MCM-48	−87	−98	−108

Table 2.
^{29}Si chemical shift of MCM-48, TPA-MCM-48 and L-arg1/TPA-MCM-48.

3.2 In vitro release study of L-arginine

3.2.1 Effect of stirring on release rate of L-arginine

In vitro release profile of L-arg_1/TPA-MCM-48 in SBF under static and stirring condition at 37°C was carried out and results are shown in **Figure 9**. Initially 28% of L-arginine is released under both the condition but after that more slower and delayed release is observed in case of static condition. It reached to 51 and 38% up to 10 h and 86 and 46% up to 32 h under stirring and static condition respectively. The slower release of L-arginine is may be due to the slow diffusion of L-arginine molecules under static condition.

As stated earlier all the experiments were carried out thrice. Statistics tool ANOVA was applied and the experimental errors within the data was found to be ±2%. The results of ANOVA single factor for release of L-arginine from TPA-MCM-48 under stirring and static condition are shown in **Table 3**. The calculated F values (0.339 under stirring and 3.062 under static conditions) are less than tabulated F value (3.09 under stirring and 3.097 under static conditions), and statistically significant difference (P) of 0.713 and 0.051 were obtained under stirring and static conditions respectively.

3.2.2 Effect of pH on release rate of L-arginine

To see the effect of pH on release rate of L-arginine, release study was carried out into SBF (pH 7.4) and SGF (pH 1.2) and results are compared (**Figure 10**). **Figure 10**. Shows slower release rate of L-arginine at lower pH. It is known that

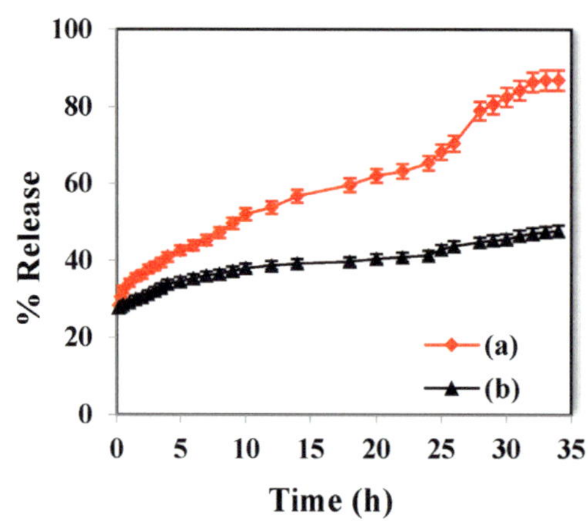

Figure 9.
In vitro release profile of L-arginine under (a) stirring condition and (b) static condition.

Condition	Source of variation	SS	df	MS	F	P-value	F crit
Stirring	Between groups	248	2	124	0.339	0.713	3.097
	Within groups	32878.21	90	365.31			
	Total	33126.21	92				
Static	Between groups	248	2	124	3.062	0.051	3.097
	Within groups	3644.169	90	40.49			
	Total	3892.169	92				

SS: sum of squares; Df: degree of freedom; MS: mean squares.

Table 3.
Results of ANOVA single factor.

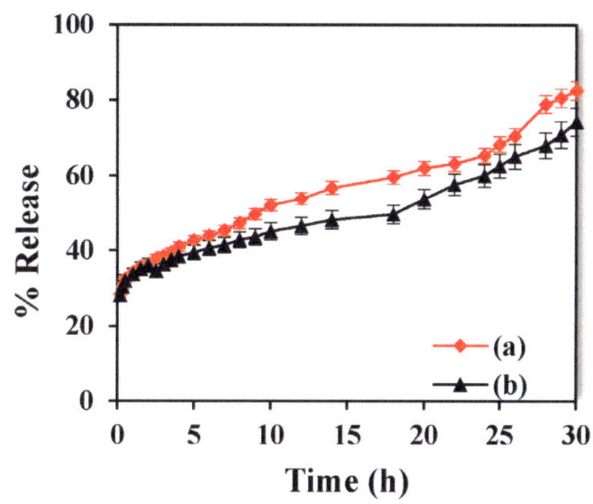

Figure 10.
In vitro release profile of L-arg1/TPA-MCM-48 in SBF (pH 7.4) and in SGF (pH 1.2).

Condition	Source of variation	SS	df	MS	F	P-value	F crit
In SBF (pH 7.4)	Between groups	248	2	124	0.339	0.713	3.097
	Within groups	32878.22	90	365.3136			
	Total	33126.22	92				
In SGF (pH 1.2)	Between groups	240.086	2	120.043	0.41457	0.661879	3.098
	Within groups	26060.44	90	289.5605			
	Total	26300.53	92				

SS: sum of squares; Df: degree of freedom; MS: mean squares.

Table 4.
Results of ANOVA single factor.

L-arginine acquires two positive charges at pH 1.2 becomes Arg^{2+} [60] which may have strong interaction with TPA-MCM-48. While at pH 7.4 it remain as Arg^{+} which may have comparatively weaker interaction with TPA-MCM-48 which allows easy diffusion of L-arginine at pH 7.4.

Under similar lines, ANOVA has also been applied for L-arginine release at different pH, and experimental error of ±2% within the data was obtained. From the results presented in **Table 4**, it can be seen that the calculated F values (0.339

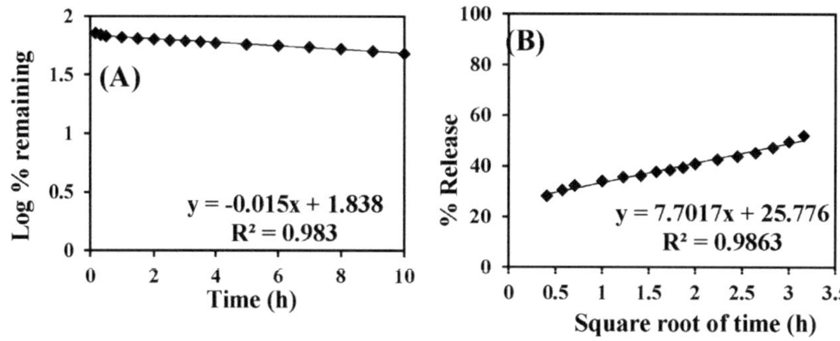

Figure 11.
(A) First order release kinetic model and (B) Higuchi model of L-arg$_1$/TPA-MCM-48.

at pH 7.4 and 0.414 at pH 1.2) are less than tabulated F value (3.09 at pH 7.4 and pH 1.2), and statistically significant difference (P) of 0.713 and 0.661 were obtained at pH 7.4 and 1.2 respectively.

3.2.3 Release kinetic and mechanism

For finding the drug release kinetic and mechanism released data up to 10 h were fitted with first order release kinetic model, Higuchi model and results are shown in **Figure 11** [61–63].

First order release kinetic model used to study the kinetic of release of soluble drugs. According to this model release of drug is concentration dependent process. **Figure 11A** shows first order release kinetic model of L-arg$_1$/TPA-MCM-48. The release of L-arginine follows the first order release kinetic model with higher linearity and co-relation coefficient (R^2 = 0.983).

Figure 11B Shows Higuchi model of L-arg$_1$/TPA-MCM-48 where % release of L-arginine was plotted against square root of time based on Fickian diffusion mechanism. According to this model release mechanism involves simultaneous penetration of SBF into the pores of carriers, dissolution of drug molecules and diffusion of these molecules from the carriers. This release data up to 10 h was best fitted with Higuchi model with higher linearity and co-relation coefficient (R^2 = 0.9863) and follows Fickian diffusion mechanism.

Hence release profile of L-arginine follows first order release kinetic model as well as Higuchi model.

3.2.4 Comparison of release profile of L-arginine from MCM-48 and TPA-MCM-48: role of TPA on release rate

To see the effect of TPA on release rate of L-arginine, release profile of L-arg$_1$/MCM-48 and L-arg$_1$/TPA-MCM-48 has been compared and results are shown in **Figure 12**. It is clear from the **Figure 12** that slower release profile is obtained for TPA-MCM-48 compared to pure MCM-48. In case of L-arg$_1$/MCM-48, initially 62% of L-arginine was released and reached to 76% up to 10 h. while in case of L-arg$_1$/TPA-MCM-48, initially 28% L-arginine released and reached to 51% up to 10 h. Thus, more controlled and ordered release rate is observed with TPA-MCM-48 systems compared to pure MCM-48. The slower release of L-arginine may be due to the strong interaction between L-arginine and terminal oxygen of TPA, which was already confirmed from FTIR spectra. Further, the pore volume of L-arg$_1$/MCM-48 is bigger than L-arg$_1$/TPA-MCM-48 (**Table 1**). So there may be easy diffusion of L-arginine molecules from MCM-48 compared to TPA-MCM-48. Here, TPA act as functionalizing agent and can hold the L-arginine for longer period of time.

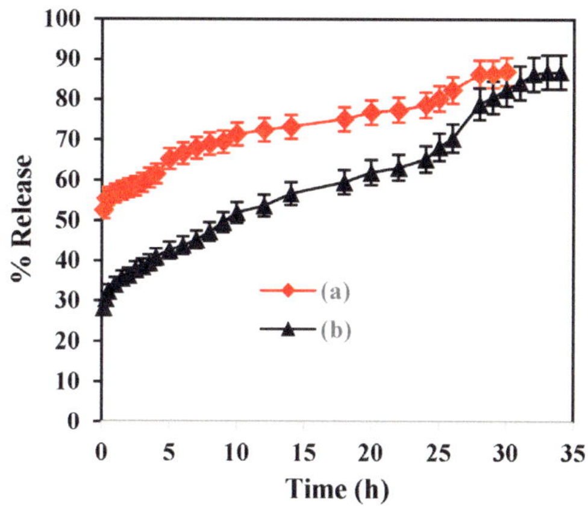

Figure 12.
Release profile of L-arg$_1$/MCM-48 and L-arg1/TPA-MCM-48.

Materials	Source of variation	SS	df	MS	F	P-value	F crit
L-arg/MCM-48	Between groups	216	2	108	1.449	0.240	3.113
	Within groups	5811.41	78	74.505			
	Total	6027.41	80				
L-arg/TPA-MCM-48	Between groups	248	2	124	0.339	0.713	3.0976
	Within groups	32878.22	90	365.3136			
	Total	33126.22	92				

SS: sum of squares; Df: degree of freedom; MS: mean squares.

Table 5.
Results of ANOVA single factor.

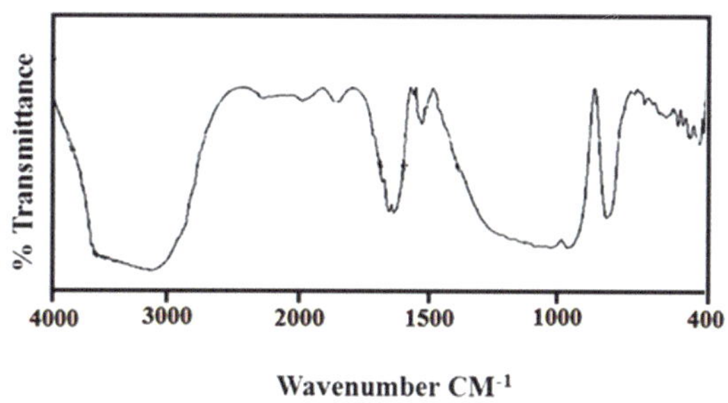

Figure 13.
FTIR spectrum of L-arg$_1$/TPA-MCM-48 after release study.

Under similar lines, ANOVA has also been applied for L-arginine release from MCM-48 as well as TPA-MCM-48, and experimental error of ±2% within the data was obtained. From the results presented in **Table 5**, it can be seen that the calculated F values (1.449 for L-arg$_1$/MCM-48 and 0.339 for L-arg$_1$/TPA-MCM-48) are lesser than tabulated F value (3.113 for L-arg/MCM-48 and 3.097 for L-arg/TPA-MCM-48), and statistically significant difference (P) of 0.240 and 0.713 were obtained for L-arg/MCM-48 and L-arg/TPA-MCM-48 respectively.

Further, to see that TPA is actually acting as functionalizing agent or not, FTIR analysis of L-arg$_1$/TPA-MCM-48 was carried out after release study and spectrum is shown in **Figure 13**. It shows that bands correspond to N-H stretching vibration and CH_3 in plan bending vibration are disappeared. This may be due to the removal of L-arginine from TPA-MCM-48 during the release study. However the bands correspond to NH_2 in plan bending vibration and C=O stretching vibration show slight shifting with lower intensity. This suggests that some amount of L-arginine was remaining inside the TPA-MCM-48. That was also confirm from release study as it shows that 86% L-arginine was release up to 32 h and then it became constant. Further, bands corresponding to TPA-MCM-48 are remaining as it is and confirm the intact structure of TPA-MCM-48.

4. Conclusion

In this chapter first time we are reporting functionalization of MCM-48 using TPA and its application as carrier for L-arginine. FTIR and NMR studies shows that L-arginine interact with TPA-MCM-48 through its N-H and C=O group. Further, BET and TEM analysis shows that structure of MCM-48 remain intact even after functionalization as well as L-arginine loading. In vitro release studies suggest that more controlled and delayed release was obtained with TPA-MCM-48 system. Presence of TPA, delayed the release rate of L-arginine and this being of great importance in the field of controlled drug delivery system. Further kinetics and mechanism study shows that release of L-arginine follows first order kinetics with Fickian diffusion mechanism.

Acknowledgements

Authors are thankful to Department of Chemistry, Faculty of Science, The Maharaja Sayajirao University of Baroda for BET and TGA analysis.

Conflict of interest

The authors declare no conflict of interest.

Author details

Anjali Uday Patel[1]* and Priyanka Dipakbhai Solanki[1,2]

1 Department of Chemistry, Faculty of Science, The Maharaja Sayajirao University of Baroda, Vadodara, Gujarat, India

2 Department of Education, Government Science College Dhanpur, Government of Gujarat, India

*Address all correspondence to: anjali.patel-chem@msubaroda.ac.in

References

[1] Rainer HB. The pharmacodynamics of L-arginine. The Journal of Nutrition. 2007;**137**:1650S-1655S. DOI: 10.1093/jn/137.6.1650S

[2] Karl IE, Klahr S, Reyes AA. Role of arginine in health and in renal disease. The American Journal of Physiology. 1994;**267**:F331-F346. DOI: 10.1152/ajprenal.1994.267.3.F331

[3] Morris SM. Arginine metabolism: Enzymology, nutrition and clinical significant. Journal of Nutrition. 2004;**134**:2743S-2747S

[4] Ann NY, Moncada S. Nitric oxide in the vasculature: Physiology and pathophysiology. Academy of Sciences. 1997;**811**:60-67. DOI: 10.1111/j.1749-6632.1997.tb51989.x

[5] Baylis C, Raij L. Glomerular actions of nitric oxide. Kidney International. 1995;**48**:20-32. DOI: 10.1038/ki.1995.262

[6] Cderbaum SD, Grody WW, Jenkinson CP. Comparative properties of arginases. Comparative Biochemistry and Physiology. Part B, Biochemistry and Molecular Biology. 1996;**114**:107-132

[7] Barbul A, Witte MB. Arginine physiology and its implication for wound healing. Wound Repair and Regeneration. 2003;**11**:419-423

[8] Marik PE, Siddiqui R, Terry C, Zaloga GP. Arginine: Mediator or modulator of sepsis? Nutrition in Clinical Practice. 2004;**19**:201-215. DOI: 10.1177/0115426504019003201

[9] Albina JE, Mills CD, Barbul A. Arginine metabolism in wounds. The American Journal of Physiology. 1988;**254**:E459-E467. DOI: 10.1152/ajpendo.1988.254.4.E459

[10] Childress B, Cowan L, Stechmiller JK. Arginine supplementation and wound healing. Nutrition in Clinical Practice. 2005;**20**:52-62. DOI: 10.1177/011542650502000152

[11] Blum A, Cannon RO, Csako G, Hathaway L, Kirby M, Mincemoyer R, et al. Oral l-arginine in patients with coronary artery disease on medical management. Circulation. 2000;**101**:2160-2164. DOI: 10.1161/01.cir.101.18.2160

[12] Fisher H, Sitren H. Nitrogen retention in rats fed on diets enriched with arginine and glycine: Improved N retention after trauma. The British Journal of Nutrition. 1977;**3**:195-208. DOI: 10.1079/bjn19770021

[13] Deng S, Dong S, Qiang G, Wanling S, Wujun X, XuYao W, et al. Amino acid adsorption on mesoporous materials: Influence of types of amino acids, modification of mesoporous materials, and solution conditions. The Journal of Physical Chemistry. B. 2008;**112**:261-2267. DOI: 10.1021/jp0763580

[14] (a)Vallet-Regi M, Balas F, Arcos D. Mesoporous materials for drug delivery. Angewandte Chemie, International Edition. 2007;**46**:7548-7558. DOI: 10.1002/anie.200604488.(b)Kmar T, Planin O. Ordered mesoporous silicates as matrices for controlled release of drugs. Acta Pharmaceutica. 2010;**60**:373-385. DOI: 10.2478/v1007-010-0037-4. (c)Manzanoab M, Vallet-Regi M. New developments in ordered mesoporous materials for drug delivery. Journal of Materials Chemistry. 2010;**20**:5593-5604. DOI: 10.1039/B922651F.(d)Yang P, Gaib S, Lin J. Functionalized mesoporous silica materials for controlled drug delivery. Chemical Society Reviews. 2012;**41**:3679-3698. DOI: 10.1039/C2CS15308D

[15] Tang Q, Chen Y, Chen J, Li J, Xu Y, Wub D, et al. Drug delivery from hydrophobic-modified mesoporous

silicas: Control via modification level and site-selective modification. Journal of Solid State Chemistry. 2010;**183**:76-83. DOI: 10.1016/j.jssc.2009.10.025

[16] Qua F, Zhua G, Lina H, Zhanga W, Suna J, Lia S, et al. A controlled release of ibuprofen by systematically tailoring the morphology of mesoporous silica materials. Journal of Solid State Chemistry. 2006;**179**:2027-2035. DOI: 10.1016/j.jssc.2006.04.002

[17] Yang P, Quan Z, Lu L, Huang S, Lin J, Fu H. Nanotechnology. 2007;**18**:235703-235715

[18] Yanga P, Quana Z, Lua L, Huanga S, Lin J. Luminescence functionalization of mesoporous silica with different morphologies and applications as drug delivery systems. Biomaterials. 2008;**29**:692-702. DOI: 10.1016/j.biomaterials.2007.10.019

[19] Lin G, Sun J, Li Z, Wang J, Ren B. Influence of different structured channels of mesoporous silicate on the controlled ibuprofen delivery. Materials Chemistry and Physics. 2012;**135**:786-797. DOI: 10.1016/j.matchemphys.2012.05.059

[20] Szegedi A, Popova M, Goshev I, Klébert S, Mihály J. Controlled drug release on amine functionalized spherical MCM-41. Journal of Solid State Chemistry. 2012;**194**:257-263. DOI: 10.1016/j.jssc.2012.05.030

[21] Kurczewska J, Lewandowski D, Olejnik A, Schroeder G, Nowak I. Double barrier as an effective method for slower delivery rate of ibuprofen. International Journal of Pharmaceutics. 2014;**472**:248-250. DOI: 10.1016/j.ijpharm.2014.06.037

[22] Qu F, Zhu G, Huang S, Li S, Qiu S. Effective controlled release of captopril by silylation of mesoporous MCM-41. ChemPhysChem. 2006;**7**:400-406. DOI: 10.1002/cphc.200500294

[23] Qu F, Zhu G, Huang S, Li S, Sun J. Controlled release of captopril by regulating the pore size and morphology of ordered mesoporous silica. Microporous and Mesoporous Materials. 2006;**92**:1-9. DOI: 10.1016/j.micromeso.2005.12.004

[24] Gai S, Yang P, Wang D, Li C, Niu N, He F, et al. Lin luminescence functionalization of MCM-48 by YVO4:Eu3+ for controlled drug delivery. Journal of RSC Advances. 2012;**2**:3281-3287. DOI: 10.1039/C2RA00862A

[25] Ambrogia V, Perioli L, Marmottinib F, Giovagnolia S, Espositoa M, Rossia C. Improvement of dissolution rate of piroxicam by inclusion into MCM-41 mesoporous silicate. European Journal of Pharmaceutical Sciences. 2007;**32**:216-222. DOI: 10.1016/j.ejps.2007.07.005

[26] Bernardos A, Aznar E, Coll C, Martínez-Mañez R, Barat JM, Marcos MD, et al. Controlled release of vitamin B2 using mesoporous materials functionalized with amine-bearing gate-like scaffoldings. Journal of Controlled Release. 2008;**131**:181-189. DOI: 10.1016/j.jconrel.2008.07.037

[27] Ambrogi V, Perioli L, Marmottini F, Moretti M, Lollini E, Rossi C. Chlorhexidine MCM-41 mucoadhesive tablets for topical use. Journal of Pharmaceutical Innovation. 2009;**4**:156-164. DOI: 10.1007/s12247-009-9073-3

[28] Vafaee M, Amini MM, Najafi F, Sadeghi O, Amani V. Modified nanoporous silicas for oral delivery of the water insoluble organotin compound: Loading and release of methylphenyltin dichloride as an anti-tumor drug model. Journal of Sol-Gel Science and Technology. 2012;**64**:411-417

[29] Maria G, Stoica A-I, Luta I, Stirbet D, Radu GL. Cephalosporin

release from functionalized MCM-41 supports interpreted by various models. Microporous and Mesoporous Materials. 2012;**162**:80-90. DOI: 10.1016/j.micromeso.2012.06.013

[30] Nastase S, Bajenaru L, Berger D, Matei C, Moisescu MG, Constantin D, et al. Mesostructured silica matrix for irinotecan delivery systems. Central European Journal of Chemistry. 2014;**12**:813-820

[31] Areana CO, Vesgaa MJ, Parrab JB, Delgado MR. Effect of amine and carboxyl functionalization of sub-micrometric MCM-41 spheres on controlled release of cisplatin. Ceramics International. 2013;**39**:7407-7414. DOI: 10.1016/j.ceramint.2013.02.084

[32] Wang Y, Sun L, Jiang T, Zhang J, Zhang C, Sun C, et al. The investigation of MCM-48-type and MCM-41-type mesoporous silica as oral solid dispersion carriers for water insoluble cilostazol. Drug Development and Industrial Pharmacy. 2013;**40**(6):819-828. DOI: 10.3109/03639045.2013.788013

[33] Kiwilsza A, Milanowski B, Druzbicki K, Coy LE, Grzeszkowiak M, Jarek M, et al. Mesoporous drug carrier systems for enhanced delivery rateof poorly water-soluble drug: Nimodipine. Journal of Porous Materials. 2015;**22**:817-829. DOI: 10.1007/s10934-015-9955-3

[34] Liu X, Che S. CFD investigation and PIV validation of flow field in a compact return diffuser under strong part-load conditions. Science China Chemistry. 2015;**58**:400-410. DOI: 10.1007/s11431-014-5743-6

[35] Nairi V, Medda L, Monduzzi M, Salis A. Adsorption and release of ampicillin antibiotic from ordered mesoporous silica. Journal of Colloid and Interface Science. 2017;**497**:217-225. DOI: 10.1016/j.jcis.2017.03.021

[36] Cuelloa NI, Elíasa VR, Mendieta SN, Longhi M, Crivelloa ME, Olivac MI, et al. Drug release profiles of modified MCM-41 with superparamagnetic behavior correlated with the employed synthesis method. Materials Science and Engineering: C. 2017;**78**:674-681. DOI: 10.1016/j.msec.2017.02.010

[37] Nastase S, Bajenaru L, Matei C, Mitran RA, Berger D. Ordered mesoporous silica and aluminosilicate-type matrix for amikacin delivery systems. Microporous and Mesoporous Materials. 2013;**182**:32-39. DOI: 10.1016/j.micromeso.2013.08.018

[38] Jesus RA, Rabelo AS, Figueiredo RT, Cides da Silva LC, Codentino IC, Fantini MCA, et al. Synthesis and application of the MCM-41 and SBA-15 as matrices for in vitro efavirenz release study. Journal of Drug Delivery Science and Technology. 2016;**31**:153-159. DOI: 10.1016/j.jddst.2015.11.008

[39] Berger D, Bajenaru L, Nastase S, Mitran R-A, Munteanu C, Matei C.Influence of structural, textural and surface properties of mesostructured silica and aluminosilicate carriers on aminoglycoside uptake and in vitro delivery. Microporous and Mesoporous Materials. 2015;**206**:150-160. DOI: 10.1016/j.micromeso.2014.12.022

[40] Arruebo M, Galan M, Navascues N, Tellez C, Marquina C, Ibarra MR. Santamari development of magnetic nanostructured silica-based materials as potential vectors for drug-delivery applications. Chemistry of Materials. 2006;**18**:1911-1919. DOI: 10.1021/cm051646z

[41] Yang P, Yang P, Teng X, Linb J, Huang L. A novel luminescent mesoporous silica/apatite composite for controlled drug release. Journal of Materials Chemistry. 2011;**21**:5505-5510. DOI: 10.1039/C0JM03878D

[42] Gai S, Yang P, Wang D, Li C, Niu N, He F, et al. Luminescence functionalization of MCM-48 by YVO4:Eu3+ for controlled drug delivery. Journal of RSC Advances. 2012;**2**:3281-3287. DOI: 10.1039/C2RA00862A

[43] Aghaei H, Nourbakhsh AA, Karbasi S, Kalbasi RJ, Rafienia M, Nourbakhsh N, et al. Investigation on bioactivity and cytotoxicity of mesoporous nano-composite MCM-48/hydroxyapatite for ibuprofen drug delivery. Ceramics International. 2014;**40**:7355-7362

[44] Popat A, Jambhrunkar S, Zhang J, Yang J, Zhang H, Meka A, et al. Programmable drug release using bioresponsive mesoporous silica nanoparticles for site-specific oral drug delivery. Chemical Communications. 2014;**50**:5547-5550. DOI: 10.1039/C4CC00620H

[45] Choi E, Lu J, Tamanoi F, Zink JI. Drug release from three-dimensional cubic mesoporous silica nanoparticles controlled by nanoimpellers. Zeitschrift für Anorganische und Allgemeine Chemie. 2014;**640**:3-4. 588-594. DOI: 10.1002/zaac.201300503

[46] Tanga Q, Xu Y, Wu D, Sun Y. Journal of Solid State Chemistry. 2006;**179**:1513-1520

[47] Hill CL, Weeks JMS, Schinazi RF. Anti-HIV-1 activity, toxicity, and stability studies of representative structural families of polyoxometalates. Journal of Medicinal Chemistry. 1990;**33**:2767-2772. DOI: 10.1021/jm00172a014

[48] Weeks MS, Hill CL, Schinazi RF. Synthesis, characterization, and anti-human immunodeficiency virus activity of water-soluble salts of polyoxotungstate anions with covalently attached organic groups. Journal of Medicinal Chemistry. 1992;**35**:1216-1221. DOI: 10.1021/jm00085a008

[49] Judd DA, Nettles JH, Nevins N, Snyder JP, Liotta DC, Tang J, et al. Polyoxometalate HIV-1 protease inhibitors. A new mode of protease inhibition. Journal of the American Chemical Society. 2001;**123**:886-897. DOI: 10.1021/ja001809e

[50] Xia L, Wang S, Feng C. Synthesis and anticancer properties of tungstosilicic polyoxometalate containing 5-fluorouracil and neodymium. Journal of Rare Earths. 2010;**28**:965-968. DOI: 10.1016/S1002-0721(09)60227-1

[51] Nomiya K, Torii H, Hasegawa T, Nemoto Y, Nomura K, Hashino K, et al. Insulin mimetic effect of a tungstate cluster. Effect of oral administration of homo-polyoxotungstates and vanadium-substituted polyoxotungstates on blood glucose level of STZ mice. Journal of Inorganic Biochemistry. 2001;**86**:657-667. DOI: 10.1016/S0162-0134(01)00233-1

[52] Yamamuro TJ. Solutions able to reproduce in vivo surface-structure changes in bioactive glass-ceramic A-W^3. Biomedical Materials Research. 1990;**24**:721

[53] Kumar D, Schumacher K, Hohenesche CDFV, Grun M, Unger KK. MCM-41, MCM-48 and related mesoporous adsorbents: Their synthesis and characterization. Colloids and Surfaces A: Physicochemical and Engineering Aspects. 2001, 2001;**109-116**:187-188. DOI: 10.1016/S0927-7757(01)00638-0

[54] Singh S, Patel A. 12-Tungsto-phosphoric acid supported on mesoporous molecular material: Synthesis, characterization and performance in biodiesel production. Journal of Cleaner Production. 2014;**72**:46-56. DOI: 10.1016/j.jclepro.2014.02.057

[55] Yijun Z, Shuijin Y. Synthesis, characterization and catalytic

application of H3PW12O40/MCM-48 in the esterification of methacrylic acid with n-butyl alcohol. Journal Wuhan University of Technology, Materials Science Edition. 2008;**23**:346-349. DOI: 10.1007/s11595-007-3346-9

[56] Kozhevnikov IV, Kloetstra KR, Sinnema A, Zandbergen HW, Bekkum HV. Study of catalysts comprising heteropoly acid $H_3PW_{12}O_{40}$ supported on MCM-41 molecular sieve and amorphous silica. Journal of Molecular Catalysis A: Chemical. 1996;**114**:287-298. DOI: 10.1016/S1381-1169(96)00328-7

[57] Kumar S, Rai SB. Spectroscopic studies of L-arginine molecule. Indian Journal of Pure and Applied Physics. 2010;**48**:251-255. Available from: http://nopr.niscair.res.in/handle/123456789/7643

[58] Okuhara T, Mizuno N, Misono M. Catalytic chemistry of heteropoly compounds. Advances in Catalysis. 1994;**41**:113-252. DOI: 10.1016/S0360-0564(08)60041-3

[59] Mizuno N, Misono M. Pore structure and surface area of $Cs_xH3_xPMo_{12}O_{40}$ (x ¼ 0e3, M ¼ W, Mo). Chemistry Letters. 1987;**16**:967-970

[60] Qiang G, Wujun X, Yao X, Dong W, Yuhan S, Deng F, et al. Interactions of lipid bilayers with supports: A coarse-grained molecular simulation study. The Journal of Physical Chemistry B. 2008;**112**:261-2267. DOI: 10.1021/jp0773051

[61] Costa P, Sousa Lobo JM. Modeling and comparison of dissolution profiles. European Journal of Pharmaceutical Sciences. 2001;**13**:123-133. DOI: 10.1016/S0928-0987(01)00095-1

[62] Singhvi G, Singh M. Review: In-vitro drug release characterization models. International Journal of Pharmaceutical Studies and Research. 2011;**2**(1):77-84

[63] Salome C, Godswill O, Ikechukwu O.Kinetics and mechanisms of drug release from swellable and non swellable matrices: A review. Research Journal of Pharmaceutical, Biological and Chemical Sciences. 2013;**4**:2,97-2,2103

Chapter 4

Polymer Functionalization of Mesoporous Silica Nanoparticles Using Controlled Radical Polymerization Techniques

Leena Nebhani, Smrutirekha Mishra and Tina Joshi

Abstract

Mesoporous silica nanoparticles (MSNs) are widely studied and are an interesting material due to its application in wide range of areas, for example, in drug delivery, catalysis, in sensors, and in adsorption and separation. Specifically, MSNs contain high surface area and large pore volume, providing high drug loading capacity, tunable pore size, surface chemistry for accommodation of a variety of guest molecules, and versatile functionalization on the external and internal surface for a broad spectrum of applications. Many new strategies have been developed for the synthesis and functionalization of mesoporous silica-based materials. The functionalization of MSNs is highly important as it leads to the development of new chemical and physical properties. Thus, preparation of these organic/inorganic hybrid structures requires facile and controlled techniques to generate enhanced properties. The grafting of polymers using controlled radical polymerization (CRP) techniques has turned out to be the best suited method to synthesize these well-defined organic-inorganic hybrid MSNs. Most common polymerization techniques are atom transfer radical polymerization (ATRP), reversible addition-fragmentation chain transfer (RAFT) polymerization, and nitroxide mediated polymerization (NMP). This chapter will be highlighting the state-of-the-art techniques for the synthesis of variety of MSNs, its functionalization using CRP techniques, and application of polymer functionalized MSNs.

Keywords: mesoporous silica nanoparticles, controlled radical polymerization, atom transfer radical polymerization, reversible addition-fragmentation chain transfer polymerization, nitroxide mediated polymerization

1. Introduction

Early innovation of mesoporous silica-based materials by Kresge and co-workers [1] in 1992 forged a strong desire in researchers to explore and demonstrate more about mesoporous silica and its novel applicability to bring tremendous changes in the field of porous materials. Porous silica network materials are broad classification [2] of inorganic materials having controllable pore size and surface area. IUPAC has classified porous materials into three categories according to its pore diameters; microporous (<2 nm), mesoporous (2–50 nm), and macroporous (>50 nm). Kresge and co-workers acknowledged a new pathway for the preparation of ordered porous MS41 family [1] by utilizing sol-gel chemistry [3] and liquid crystal micellar templating route [4]. Further utilizing this concept, different porous structures were prepared such as MCM-41 (hexagonal) [5] and MCM-48 (cubic) [6]. Since then, mesoporous silica has been explored by different groups such as SBA [7] series from University of California Santa Barbara by Stucky group, KIT [8] series by Korean researchers, and FDU [9] series from Fudan University China by Prof. Zhao's group. Despite many synthesis strategies, mesoporous silica cannot be used in certain applications due to its amorphous pore wall structure [10]. Hence, it is enormously required to find new pathways to modify amorphous surface of mesoporous silica with variety of organic groups or polymers, in order to find its suitability for diverse applications.

The main motivation of this chapter is to highlight different strategies for construction of MSNs and essential techniques for the characterization of MSNs. The functionalization of MSNs by polymer grafting using controlled radical polymerization (CRP) [11] techniques and its implementation in distinct domains has been discussed in detail.

1.1 Origination of different types of MSNs

Prior discovery of mesoporous materials in 1970s did not make any tremendous change in the Materials Research until 1992, when Mobil Researchers (Kresge and co-workers) reported mesoporous molecular sieves using liquid crystalline template. These materials were designated as Mobil Composition of Matter (MCM) or Mobil Crystalline Materials. Generally, these MCM materials are synthesized utilizing different cationic surfactant-based templating agents. MCM are classified into different types such as; MCM-41 [12], MCM-48 [13], and MCM-50 [13]. Out of all, MCM-41 is one of the most investigated materials from Mobil crystalline materials family for various applications. MCM-41 contains a regular pore geometry of 2.5–6 nm with hexagonal type of ordered porous structure. These pore diameters can be tuned by varying the surfactants used. Other MCM materials are synthesized by varying the surfactant and its loading leading to cubic arrangement for MCM-48 and lamella-type arrangement for MCM-50 as shown in **Figure 1** [14].

Other than cationic surfactants, nonionic copolymer-based surfactants are also used in the synthesis of Santa Barbara Amorphous (SBA) type of MSNs. Copolymers of poly(ethylene oxide) and poly(propylene oxide) are commonly used as templating agents for synthesis of SBA. Once again in case of SBA materials, based on the type of copolymers used, it can be synthesized into different ordered patterns such as cubic structured for SBA-11, 3-D hexagonal structured for SBA-12,

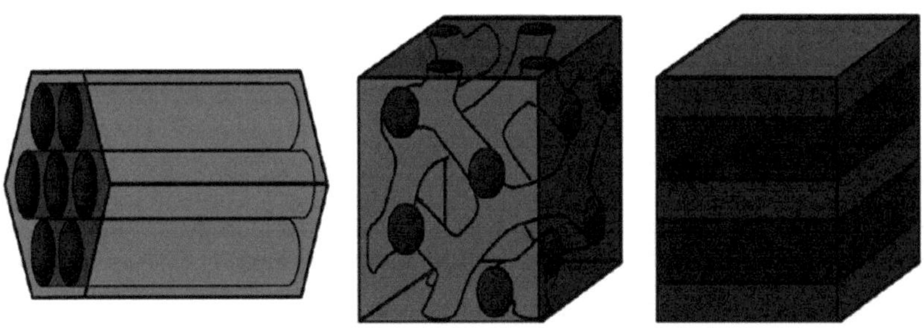

Figure 1.
Different types of pore geometry structure for MCM family (left side for MCM-41 with hexagonal structure, center for MCM-48 with cubic structure, and right for MCM-50 with lamella structure). Reproduced with permission from Ref. [14]).

hexagonal for SBA-15 [15], and 3-D cubic cage-structured for SBA-16 [16]. By altering the ratio of propylene oxide to ethylene oxide in the copolymer, different ordered structures can be obtained for this SBA. Mostly, highly ordered porous network of SBA-15 has been commonly used for different biomedical applications. These materials are broadly different than MCM due to its thicker amorphous wall and larger pore diameter of 4.5–30 nm.

Another interesting MSNs family material was discovered by Japanese researchers named as FSM-16 [17] (folded sheets of mesoporous materials) synthesized by an ion-exchange process between cations of layered structured sodium silicate and surfactant followed by hydrothermal treatment. Based on these investigations, numerous MSNs were then synthesized and designated such as Technical Delft University (TUD-1) [18], Korean Institute of Technology (KIT) [8], Hiroshima Mesoporous Material (HMM-33) [19], Indian Institute of Technology Madras (IITM), and Centrum voor Oppervlaktechemie en Katalyse/Centre for Research Chemistry and Catalysis (COK-12), [20] according to its pore size and symmetry. The detailed pore symmetry and pore size for different types of MSNs has been discussed in **Table 1**.

1.2 Synthesis and characterization of MSNs

The synthesis of MSNs is commonly conducted in the presence of a templating agent, also called as structure-directing agent. The two important types of templating agents commonly used are soft templates, for example, cationic, anionic and nonionic surfactants, block co-polymers, etc. The other type of templating agent used is, for example, preformed mesoporous silica or carbon. The commonly used precursor for silica formation is tetraethoxysilane (TEOS); however, other inorganic salts can also be used. The synthesis of MSNs can be performed using bases, for example, sodium hydroxide, potassium hydroxide, and ammonium hydroxide, as well as under acidic conditions. Finally, the mesoporous structure is achieved by the extraction of templates via various methods, for example, calcination and solvent extraction. The mechanism of the formation of mesoporous silica has been studied using combination of several techniques, for example, X-ray diffraction (XRD) [21], small angle X-ray scattering (SAXS) [22], solid state NMR [23], electron microscopy, thermal analysis,

Type of MSNs	Pore symmetry	Pore diameter (nm)	Type of surfactant
MCM-41	Hexagonal	1–8	Cationic (CTAB)
MCM-48	Cubic	2–5	Cationic (CTAB)
MCM-50	Lamellar	2–5	Cationic (CTAB)
SBA-11	Cubic	2–5	Non-ionic co-polymer (Igepal CO630)
SBA-12	Hexagonal	3–5	Non-ionic co-polymer (Brij-76)
SBA-15	Hexagonal	6–15	Non-ionic co-polymer (Pluronic P123)
SBA-16	Cubic	5–15	Non-ionic co-polymer (Pluronic F127)
KIT-5	Cubic	4–10	Non-ionic co-polymer (Pluronic F127)
COK-12	Hexagonal	4–15	Non-ionic co-polymer (Pluronic P123)
FSM-16	Sheet	2–5	Cationic (CTAB)
TUD-1	Foam	2–25	Cationic (CTAB)

Table 1.
Detailed inherent characteristics of different types of MSNs.

and sorption analysis [24]. For the preparation of MCM41/Mobil, scientists proposed that it is a liquid-crystal template mechanism because liquid crystals of surfactants work as templates to generate mesopores. The mesopores that are created after extraction have a long range order which can be analyzed by X-ray diffraction (XRD) at low 2θ (small angle XRD, 2θ = 2–10). The wide angle XRD is commonly used to study inorganic amorphous pore walls, which are formed by the condensation of TEOS (wide peak from 2θ =20–25). The diffraction peaks obtained by performing XRD on MSNs can be used to identify a known meso-structure. In the case of MCM-41, four different diffraction peaks have 1/d value in the ratio of $1:\sqrt{3}:\sqrt{4}:\sqrt{7}$, which corresponds to 2D hexagonal pore structure and space group of P6mm symmetry.

In order to monitor the kinetics and growth mechanism of MSNs, SAXS [25] has been utilized. SAXS is an important and nondestructive technique for the structural characterization of MSNs in solution which is not possible with other techniques like transmission electron microscopy (TEM). Using SAXS, MSNs can be utilized for studying their morphology and dispersion in any form such as in colloids, nanopowder, nanocomposite, and microemulsions. Other than predicting the structure and morphology of the MSNs, one can understand the growth mechanism during synthesis of MSNs by synchrotron time-resolved small angle X-ray scattering. Yi et al. [26] revealed a better understanding for the growth mechanism of MSNs by predicting the scattering proportions of the incident X-ray beam via synchrotron time-resolved small angle X-ray scattering measurements. This technique was modeled at the micellar level upon addition of CTAB to predict the shape and size of micelles at the interface. It was concluded that the micellar size changes from ellipsoid to spherical upon addition of TEOS due to its solubilization in the medium. Hence, it was efficiently demonstrated that utilizing SAXS, a good understanding of growth kinetics of MSNs can be obtained.

Electron microscopy [24, 27], mainly high resolution TEM, can be utilized for the structural characterization of MSNs. An electron microscopy technique uses an electron beam accelerated at a high voltage as the source of radiation. The back-scattered and secondary electron provides structural characterization, while X-ray generated provides chemical composition. As shown in **Figure 2**, mesoporous structure can be directly observed under TEM. The white area represents the pores and black area represents amorphous silica framework.

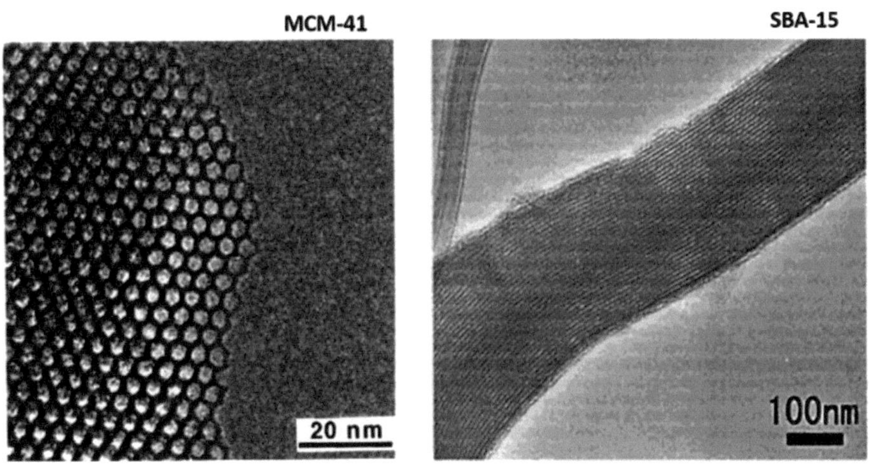

Figure 2.
TEM micrographs for (a) MCM-41 (reproduced with permission from Ref. [24]) and (b) SBA-15 (reproduced with permission from Ref. [27]).

MSNs are commonly characterized for surface area, pore volume, pore size, and pore size distribution using physical adsorption of an inert gas, for example, nitrogen. As the pore size in the mesoporous materials is greater than 2 nm and less than 50 nm, commonly type IV and V isotherms are obtained. The surface area is calculated using Brunauer-Emmett-Teller (BET) equation. Many methods are available for calculation of pore size and pore size distribution, for example, Barrett, Joyner, and Halenda (BJH) method, non-local density functional theory (NLDFT), etc.

2. Polymer functionalization of MSNs

The grafting of polymers from the surface of MSNs is one of the efficient techniques for the surface functionalization of MSNs. The growth of polymeric chains from MSNs is highly important in the area of Materials Science, as it provides MSNs with the properties which are desired in several applications [28].

The commonly preferred grafting techniques are based on different surface initiated polymerization methods such as reversible addition-fragmentation chain transfer polymerization (RAFT), atom transfer radical polymerization (ATRP), and nitroxide-mediated radical polymerization (NMP). The polymer functionalized MSNs can be easily used in the field of drug delivery, catalysis, and sensors to name a few. As shown in the **Figure 3**, the polymer chains can be physically adsorbed on MSNs (physisorption) or can be covalently attached (chemisorption). Generally, the physisorption method is not an appropriate technique since it is noncovalent in nature and

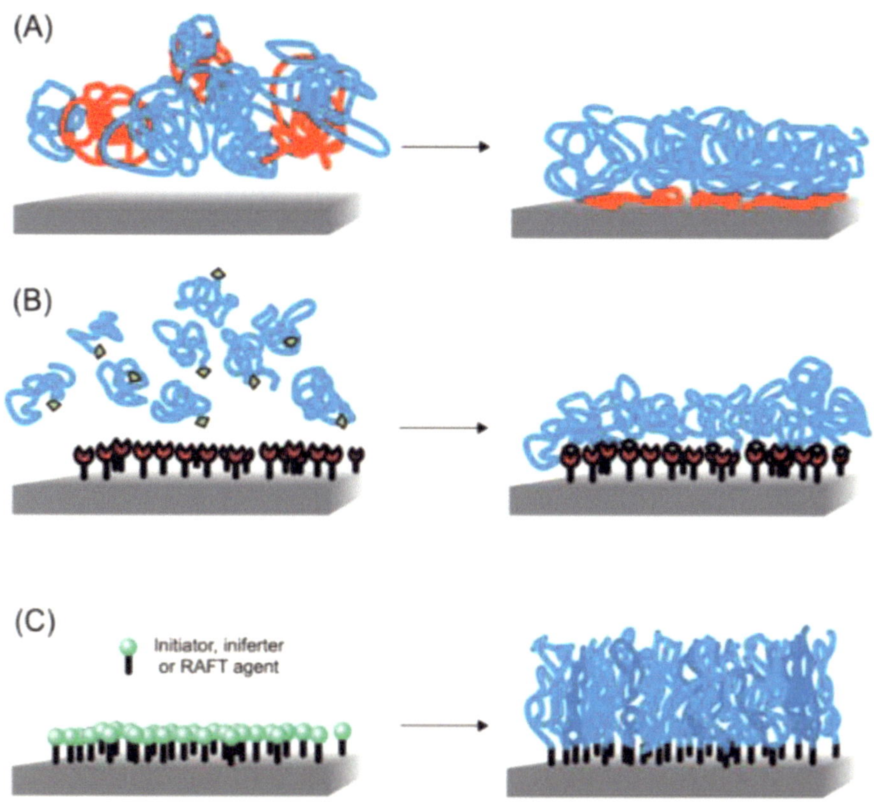

Figure 3.
Approaches for grafting of polymer chains from solid substrate. (A) Physical adsorption, (B) Grafting-to approach, and (C) Grafting-from approach (reproduced with permission from Ref. [28]).

is reversible. The chemical grafting via covalent bonding is a favored method due to its irreversible compatibility in the midst of organic and inorganic group. The chemical grafting can be carried out by two means such as "grafting-to" and "grafting-from." The grafting-to utilizes covalent bonding by reacting pre-synthesized macromolecules carrying specific group with specific group attached on the inorganic surface, while grafting-from involves modifying the surface of MSNs with an initiator followed by surface initiated polymerization. The growth of polymer chain can be achieved by controlled radical polymerization techniques, for example, RAFT, ATRP or NMP.

2.1 RAFT polymerization

In the year 1998, RAFT polymerization was first reported by Rizzardo et al. [29], involving an array of addition-fragmentation equilibria mechanism as shown in the **Figure 4**. The polymerization is started by activation of the initiator generating radical. This radical then drives toward the thiocarbonylthio group of the RAFT agent which then forms a carbon-centered intermediate by reacting with the propagating radicals. Once again the intermediate undergoes a β scission generating radicals in order to reinitiate propagation. The process will continue by adding reversibly to the chain transfer agent. Finally, equilibria point is generated between the dormant and propagating species.

This polymerization demands a convenient transfer agent having high transfer constant in the radical polymerization for any monomer under the required polymerization conditions. RAFT polymerization is a highly effective technique than ATRP and NMP, since it can be easily carried out in different reaction conditions and is suitable for wide range of monomers. Since the polymer chains in this method carries a thiocarbonyl group, the synthesis of any block copolymer using a second monomer is also convenient using RAFT polymerization. Thus by utilizing this technique, different inorganic surfaces can be functionalized with the polymeric chains. Ma et al. [30] reported grafting of copolymers of positively charged quaternary amines and polyethylene glycol (PEG) via RAFT polymerization using grafting-from technique from the surface of MSNs. The schematic representation for this synthesis is detailed in **Figure 5**. Initially, MSNs are synthesized, which are later on modified with amine functionalities via post-modification. Then, trithiocarbonate-based RAFT agent was synthesized from the surface of amine functionalized MSNs. Subsequently, these RAFT functionalized MSNs are utilized for RAFT polymerization for grafting of poly(ethylene glycol) and 2-(dimethylamino) ethyl acrylate. The successfully synthesized co-polymer-grafted MSNs were later on reacted with methyl iodide (MeI) in order to generate quaternary amines from tertiary amines. The aim of introducing quaternary amines was to maintain the positive charges on the surface of MSNs even in different pH conditions and enable the dispersity of the particles due to its positive charge leading to electrostatic repulsion. Finally, these co-polymer-grafted MSNs were applied for in-vivo studies in drug-delivery application.

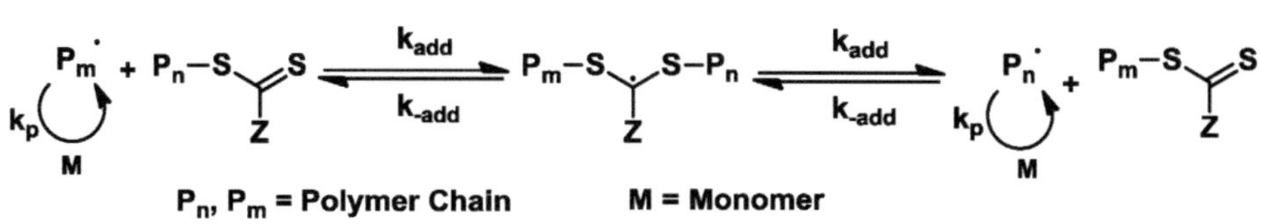

Figure 4.
Addition-fragmentation equilibria during RAFT polymerization.

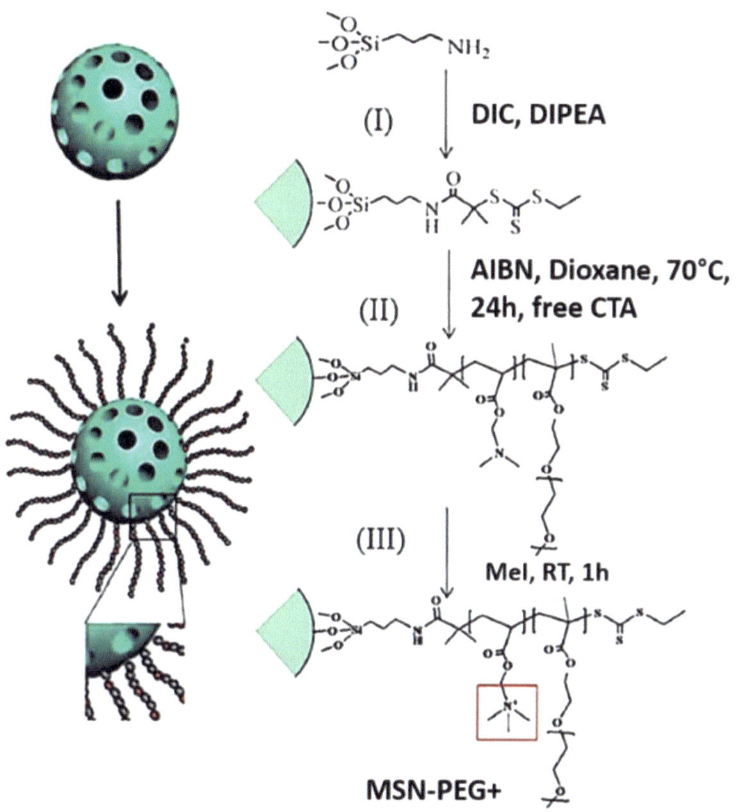

Figure 5.
Schematic representation for synthesis of MSNs grafted with copolymer of positively charged quaternary amines and PEG via RAFT as reported by Ma et al. (reproduced with permission from Ref. [30]).

Mishra et al. [31, 32] described grafting of poly(*N*-isopropyl acrylamide) (PNIPAM) onto MSNs via surface-initiated RAFT polymerization. In this work, initially two different R group containing organoalkoxysilane-based RAFT agent were synthesized. Then, these organoalkoxysilane-based RAFT agents were functionalized onto MSNs via co-condensation resulting in-built RAFT agent containing MSNs. Subsequently, NIPAM was polymerized further from these in-built RAFT agent containing MSNs via surface-initiated RAFT polymerization as shown in **Figure 6**. Further to understand about the thermoresponsiveness behavior of MSNs grafted with PNIPAM so that it can be utilized in drug delivery application, in-vitro studies of fluorescein dye and doxorubicin drug was investigated at different temperatures. Hong et al. [33] studied post-modification of the surface of MSNs by a trithiocarbonate-based RAFT agent. MSNs were synthesized without CTAB extraction and then it was post-modified by 5,6-epoxyhexyltriethoxysilane (EHTES) on the outer surface of MSNs as shown in **Figure 7**.

Further, EHTES functionalized MSNs were refluxed in methanol/hydrochloric solution in order to extract CTAB, as well as epoxyhexyl group were converted to 5,6-dihydroxyhexyl units. Then, trithiocarbonate-based RAFT agent was functionalized to MSNs via esterification. Thus, adopting such mechanism, polymers such as polystyrene, poly(acrylic acid), and PNIPAM were grafted from the surface of MSNs via surface-initiated RAFT polymerization. Gonçalves et al. [34] reported a pH-responsive polymer coated on surface of MSNs via RAFT polymerization. Nanocontainers based on fluorescent core

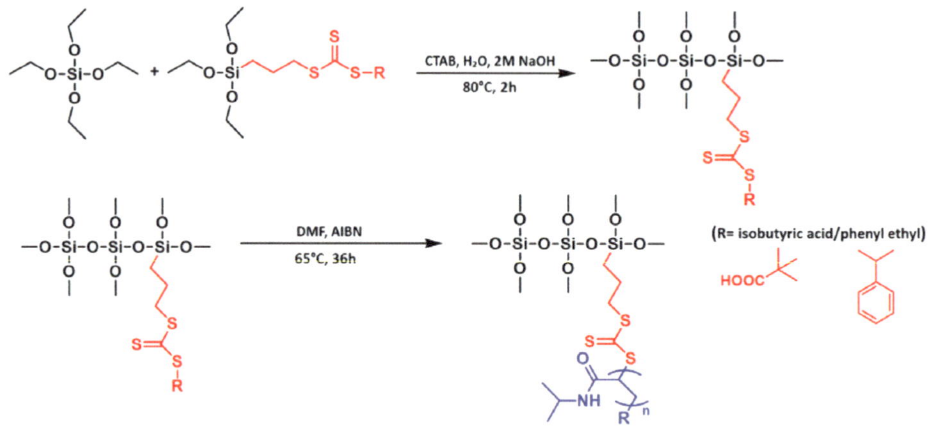

Figure 6.
Schematic representation for MSNs functionalized with different R group containing organoalkoxysilane RAFT agent via co-condensation and further grafting with PNIPAM via surface-initiated RAFT polymerization as reported by Mishra et al. (reproduced with permission from Ref. [31]).

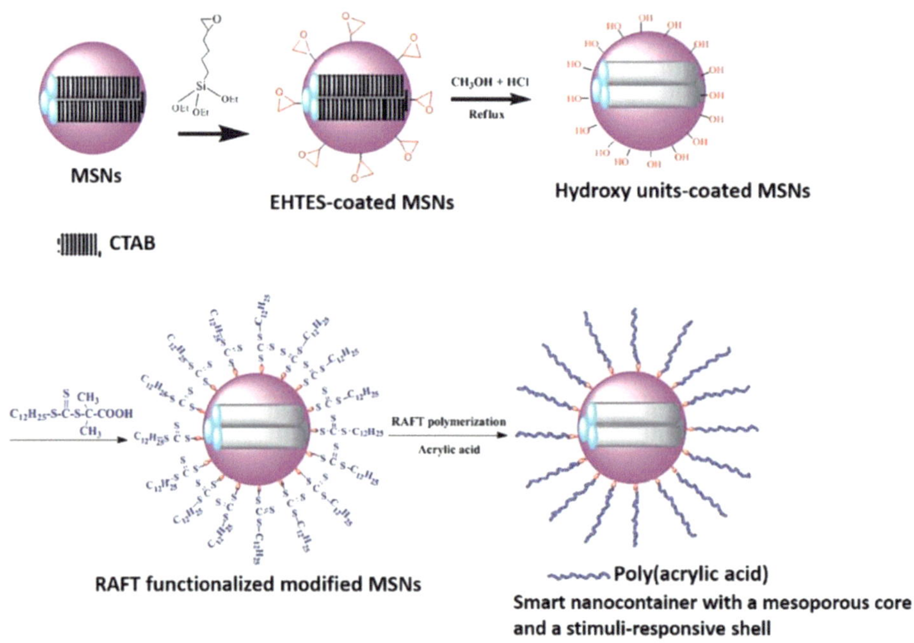

Figure 7.
Schematic representation for grafting of poly(acrylic acid) on the exterior surface of MSNs via RAFT polymerization as reported by Hong et al. (reproduced with permission from Ref. [33]).

MSNs were prepared, which were coated with a shell of pH-responsive polymer poly(2-(diisopropylamino)ethyl methacrylate). The synthesis was carried out by incorporation of a fluorescent dye in the silica network and by functionalization of a RAFT agent on the outer surface of MSNs. Further utilizing this RAFT polymerization, the pH-responsive polymer was coated only outside the surface of MSNs in order to keep the pores free to encapsulate drug into it. The pH-responsive polymer-grafted MSNs displayed an outstanding drug release from the pores at low pH due to electrostatic interaction of the cargo with the charged silica and due to the pH-responsive polymer.

2.2 ATRP and NMP polymerization

Among other methods of controlled radical polymerization, surface-initiated atom transfer radical polymerization (ATRP) is gaining importance for the synthesis of polymer on the mesoporous surface of inorganic moieties. ATRP is an another technique through which growth of polymeric chains of controlled molecular weight can be performed on the surface of MSNs [1]. The general mechanism of ATRP as shown in **Figure 8** comprises of an alkyl halide initiator, R–X, or a latent polymer chain, P_n–X, (where X is a halogen), which is activated by a metal/ligand catalytic complex in a lower oxidation state, for example, [Cu(I)/L], resulting in the formation of a propagating radical and the catalytic complex in a higher oxidation state [X–Cu(II)/L]. Then, monomer units add to the growing radical and polymerization occurs.

Since MSNs provide high specific area and thermal stability, they have been further modified so as to achieve enhancement in the performance which is required in most of the applications. In field of biomedicine, Huang et al. [35] discussed the applications of visible light-induced surface-initiated ATRP for drug delivery using mesoporous silica polymer nanocomposites. They synthesized a nanocomposite named MSNs-NH_2-poly(IA-co-PEGMA) using itaconic acid (IA) and poly(ethylene glycol)methyl acrylate (PEGMA), which were grafted on MSNs. Importantly, they have loaded an anticancer drug cisplatin effectively onto these nanohybrid structures and thus performed studies based on controlled drug release in response to a change in pH. The light-induced surface-initiated ATRP proved promising compared to traditional ATRP, as the prior could avoid the use of metallic catalyst and other organic ligands as well as the reaction can be performed at much lower temperature and with increased rate of polymerization. **Figure 9** shows the preparation of the nanohybrid along with the control released studies performed for cisplatin.

In recent times, ATRP with activators regenerated by electron transfer (ARGET) is popular among researchers. It provides major improvements in usage as it can be performed easily using closed vial and involves low concentration of copper catalyst. In ARGET ATRP, a reducing agent (tin(II) 2-ethylhexanoate, or vitamin C) is added to convert Cu(II) to Cu(I) and to remove the oxygen. ARGET ATRP has been successfully utilized for growing polymer brushes on flat surfaces as well as on nanoparticles. Cao and Kruk [9] demonstrated ARGET-ATRP for modification of SBA-15 with different polymers, for example, polystyrene and poly(methyl methacrylate) of low polydispersity index. The technique of ARGET ATRP was demonstrated to be useful for the synthesis of high surface area silica-polymer composites. The work concluded the simplicity of the technique for synthesizing well-defined polymer brushes in porous architecture of silica. **Figure 10** illustrates the synthesis route of grafting polymer in nanopores of MSNs.

Mesostructure cellular foam materials (MCFs) have large pore of 15–40 nm and can easily host large molecules. Zhou et al. [36] reported the synthesis and design engineering of drug system based on thermoresponsive polymer (PNIPAM) inside

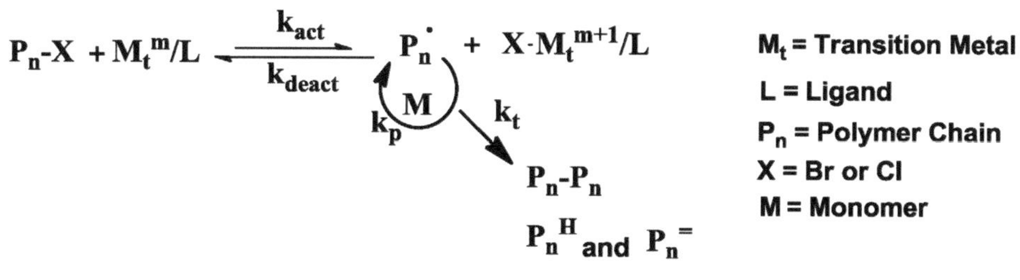

Figure 8.
Activation-deactivation during ATRP.

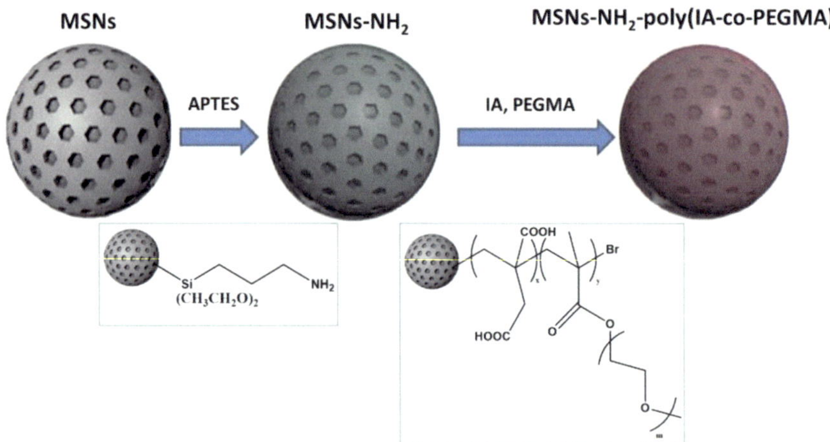

Figure 9.
Synthesis route of MSNs-NH$_2$-poly(IA-co-PEGMA) via a catalyst free ATRP (reproduced with permission from Ref. [35]).

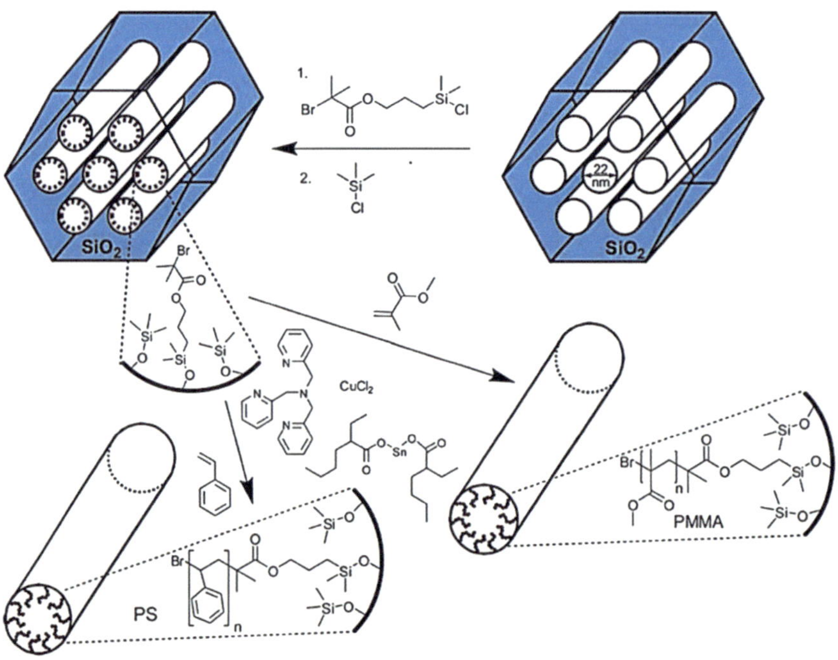

Figure 10.
Grafting poly(methyl methacrylate) and polystyrene in the cylindrical mesopores of SBA-15 silica using ATRP (reproduced with permission from Ref. [9]).

MCFs using ATRP. The controlled release of the drug inside MCFs was investigated via thermoresponsive nature of the polymer used under environmental temperature. The occurrence of polymerization was restricted to the surface of MCFs. They have investigated the high storage capacity of 58 wt.% for ibuprofen (IBU/silica) which was comparatively much higher than that reported for functional SBA-15 (37 wt.%). Zhou et al. [37] demonstrated the synthesis of PGMA (poly(glycidyl methacrylate)) MSNs based nanocarriers capable of drug delivery applications using disulfide bonds as redox-responsive cross-linkers. The successful grafting of PGMA on the surface of MSNs was performed using SI-ATRP as shown in **Figure 11**. The epoxy ring opening

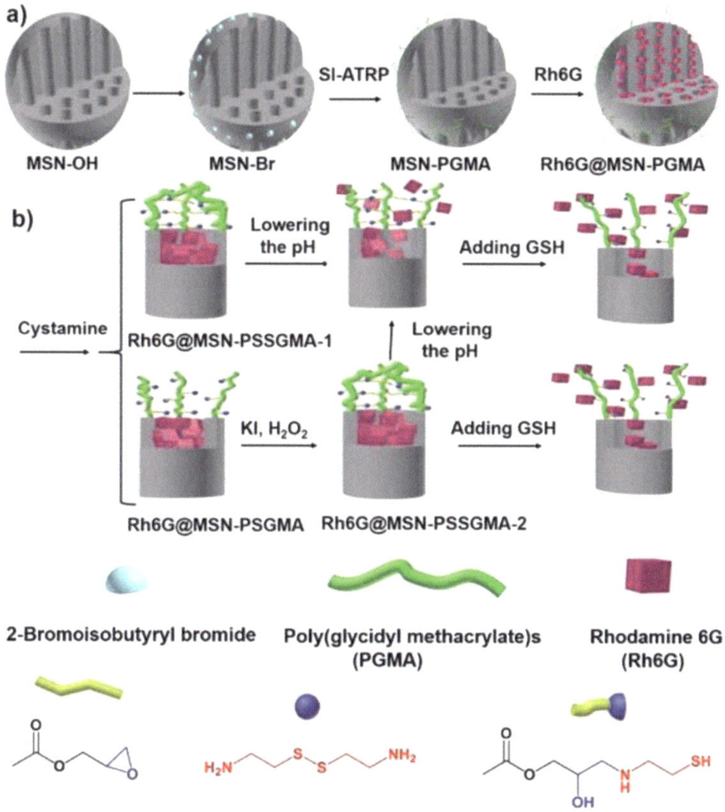

Figure 11.
Schematic illustration of (a) the synthetic route of MSN–PGMA and Rh6G (Rhodamine 6G)-loading process; (b) cystamine cross-linked PGMA network on MSNs and its Rh6G release in response to pH and GSH (reproduced with permission from Ref. [37]).

reactions of PGMAs was performed using cystamine dihydrochloride and a facile mechanism for building disulfide-containing cross-linked structures. The dual external stimuli responsive system based on pH and GSH (glutathione) in cancer cells. These biocompatible materials having degradable S–S bonds can find application in biomedical area. Hence, ATRP is a powerful approach and can be effectively used for grafting of polymers or polymer brushes from mesoporous surfaces. ATRP allows to achieve grafting of polymer chains of controlled molecular weight and low polydispersity index while maintaining the pore accessibility and moderate to high surface area.

Nitroxide-mediated polymerization (NMP) is an another easy way of controlled radical polymerization, which gives well-defined, functional macromolecular structures. IUPAC has endorsed the term "nitroxide" to "aminoxide" and thus recommended aminoxyl-mediated radical polymerization (AMRP). NMP involves the process of activation/deactivation having a reversible combination of propagating radicals with free nitroxide radicals as shown in **Figure 12**.

A variety of nitroxide and nitroxide-based alkoxyamines have been reported in the literature in order to conduct polymerization of different monomers. Adequate surface-active initiators required for the development of grafting on the inorganic surface via NMP is much less explored. The nitroxide radical combines with carbon-centered radicals and gives alkoxyamines. Due to thermal activation requirement, NMP is considered as a simple and effective method for polymerization. In NMP, no further purifications are required except a precipitation to remove unreacted monomer. The materials prepared using NMP

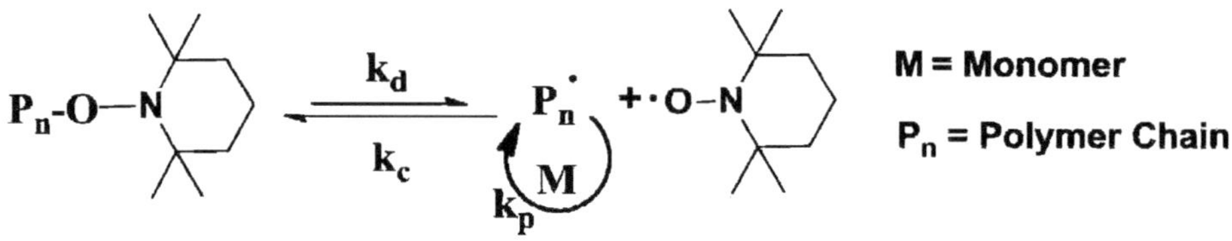

Figure 12.
Activation-deactivation during NMP.

can find applications in various fields such as optoelectronics, nanoporous, nanostructured materials, and surfactants/dispersants, etc. [38]. Charleux [39] reported the polymerization of styrene using NMP on the surface of various types of ordered mesoporous silica (OMS). NMP, being a reversible termination technique, can provide surface-initiated polymerization mechanisms for mesoporous substrates. This was able to successfully determine that there is no effect of the OMS particle structure on the polymerization kinetics. The important conclusions drawn from the experiments are that pore size of 5 nm is adequate for diffusion reactions while 2 nm are too small for the same. In addition, large pore size (≥5 nm) is important for the pore connectivity. Barthalome et al. [40] reported the two-step synthesis of nitroxide-mediated radical polymerization of styrene onto the silica nanoparticles. Using this technique, it was manifested that polymer graft densities of around 110 chains per particle were obtained. The alkoxyamine-functionalized silica nanoparticles were prepared, followed by growth of polystyrene chains having narrow polydispersity and targeted molecular weights. By varying styrene-to-initiator molar ratios, an insight of the growth mechanism was obtained by comparison of molecular weights and polydispersity index of the grafted and free polymeric chains. The novelty of the approach was found in using triethoxysilyl-functionalized alkoxyamine initiator. Using this approach, the well-defined structures of polystyrene-coated MSNs were obtained, which can find good potential as nanocomposite in the field of materials science.

3. Applications of polymer functionalized MSNs

Mesoporous silica nanoparticles find versatile potential in the developing field of catalysis [41], coating, sensing, and as drug carriers [42]. Due to its high surface area, bio-compatibility, chemical stability, and tunable porous architecture, MSNs are capable of various modifications which can be easily anchored on its surface [43]. Accordingly, modification/functionalization methods, e.g., co-condensation method, encapsulation process, and post-synthesis like grafting-to and grafting-from techniques are recently captivating immense attentions. Due to its low toxicity and high drug loading capacity, MSNs are now extensively investigated as smart system which can be used for controlled and targeted drug-delivery systems. Targeting ligands toward the selective diseased area have proven to be a difficult task. The concept of a stimuli-responsive drug release system takes into account of control and targeted release of the required dosage of drug [44]. Internal as well as external stimuli such as pH, temperature, light, magnetic fields, electric fields, ultrasounds, and redox potential can be used in many cases. Paris et al. [45] demonstrated a new approach to stimuli-responsive system by making use of hybrid MSNs. These hybrid-MSNs are mesoporous silica as carriers coated with dual temperature and ultrasound-responsive copolymer having lower critical solution temperature (LCST) below 37°C which act as nanogates sensitive to ultrasound stimulations. On providing ultrasound, copolymer changes its physical state at the respective temperature leading to release of loaded molecules from carriers. Doxorubicin-loaded hybrid MSNs were incubated with LNCaP (androgen-sensitive human prostate adenocarcinoma) cells to show their capacity to induce cell death only when this hybrid system was exposed to ultrasound as shown in **Figure 13**.

Stimuli-responsive polymers such as poly(methacrylic acid) (PMAA) and poly(*N*-isopropyl acrylamide) (PNIPAM) have been widely investigated. Zheng et al. [46] has reported thermo- and pH dual-responsive nanocarrier silica acting as core and block copolymer as shell. Surface-initiated-RAFT (SI-RAFT) polymerization of methacrylic acid (MAA) and *N*-isopropyl acrylamide (NIPAM) was performed on the silica surface. The polymer-grafted MSNs showed higher loading efficiency of drug doxorubicin (DOX) and these dual-responsive nanoparticles were used as a drug carrier. The rate of drug release depends upon both temperature and pH of the surrounding media. The agglomeration was observed under acidic and elevated temperature conditions, whereas particles were dispersed in an aqueous media at high pH and low temperatures. The features of resulting SiO_2-PMAA-b-PNIPAM show potential for therapeutic applications as revealed by a cellular uptake study. These systems were able to deliver DOX successfully into the nuclei of HeLa cells (cervical cancer cells from Henrietta Lacks). In addition, cytotoxicity study revealed that DOX-loaded nanoparticles were more active and effective with HeLa cells than free equivalent dosage of DOX. Liu et al. [47] reported dual-responsive smart polymer coated on MSNs for laryngeal carcinoma therapy. Thermo/pH dual-responsive polymer poly[(*N*-isopropyl acrylamide)-co-(methacrylic acid)] was grafted onto mesoporous silica to act as a "valve" to restrain the diffusion of the incorporated drugs in and out of the pore channels. They reported "on-off" drug release pattern at higher temperature and lower pH value. The folic acid molecules were further attached to the polymer-modified MSNs in order to target Hep2 cells. The synthesis of polymer-grafted MSNs and control delivery of drugs is depicted in **Figure 14**.

The organic/inorganic nanocomposites possess combined advantages as inorganic materials are thermally stable and give rigidity, whereas organic materials are much more flexible and can be easily processed. Also, the size of fillers gives increased interfacial area than ordinary composites. Silica is considered as suitable

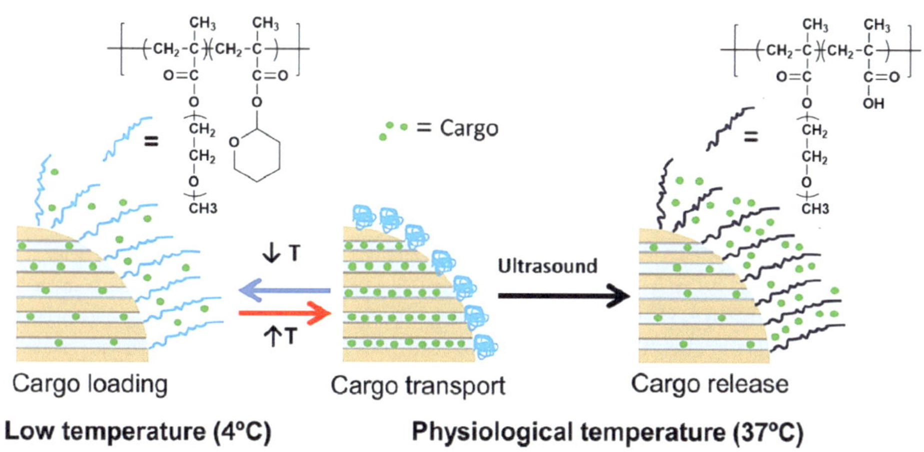

Figure 13.
Illustration of the behavior of dual-responsive system in aqueous medium (reproduced with permission from Ref. [45]).

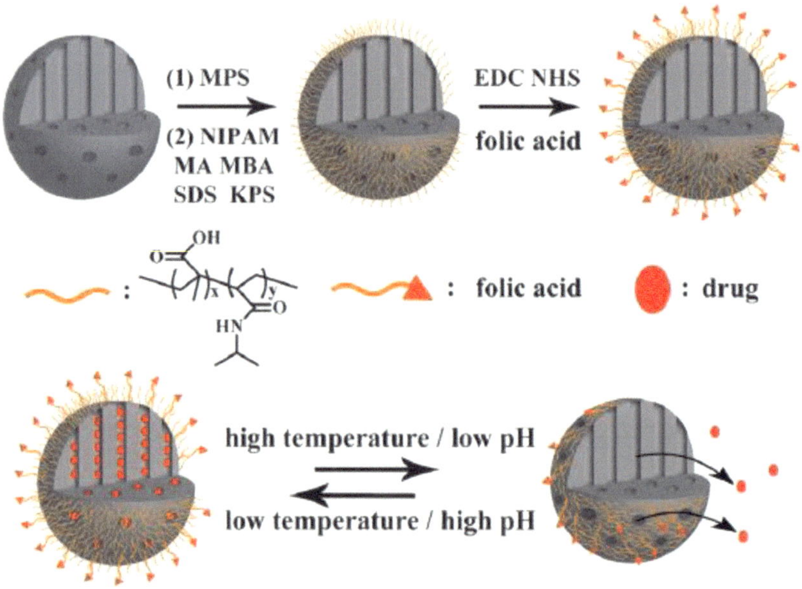

Figure 14.
Illustration of synthesis of dual-responsive drug delivery system (reproduced with permission from Ref. [47]).

nanofillers for such nanocomposites and MSNs find its unique application in catalysis. Tang et al. [48] synthesized a series of nanocatalyst based on MSNs for polymerization of 1,3-butadiene and simultaneously immobilizing salicylaldimine cobalt complexes on MSNs surface. Combining MSNs with methylaluminoxane (MAO) gives excellent catalytic efficiency for the polymerization. The yield of the polymer and its molecular weight also depends on the size of the particles used for catalysis. Thus, this catalysis process provides an easy way to directly synthesize polymer/silica composites. Further, Tang et al. [49] demonstrated a novel synthesis route for in situ polymerization of isoprene wherein MSNs were used as catalyst to synthesize rare earth oxide-containing luminescent polymer/silica nanocomposites having core-shell architecture. In this case, MSNs functioned both as catalyst as well as inorganic core for nanocomposites and caused phase separation for more interfacial interactions. The mesoporous materials have shown its potentiality over microporous zeolites in the area of heterogeneous catalysis. Keeping note, Albuquerque et al. [50] reported new improved catalysts for trans-esterification of vegetable oils with methanol using MSNs such as SBA-15, MCM-41, and fumed silica. A linear model was reported and found to predict the conversion rate for the product formation. The reported process states that the most active catalyst consisted with a sample having 14 wt.% of CaO on SBA-15, achieving 95% of conversion with sunflower oil in 5 h and 65% with castor oil in 1 h. Similar studies were reported by Abdullah et al. [51] by using responsive surface methodology for trans-esterification of palm oil using mesoporous K/SBA-15. The type of alcohol and oil used with various reaction time and temperature, their molar concentration along with catalyst type all were the factors affecting the efficiency for production of biodiesel. Potassium (K) was introduced via conventional impregnation method to make SBA-15 act as basic catalyst sites. An effective mathematical method was developed which can be used to determine the product yield (biodiesel) with optimum reaction conditions with the accuracy. The relation between reaction time, temperature, and catalyst loading was successfully elucidated. It was reported that 87.3% was highest yield, which was obtained when the system conditions were 11.6 mol/mol for methanol to oil ratio with 3.91 wt.% of catalyst with reaction time of 5 h at 70°C reaction temperature. Farjadian et al. [52] reported comparative studies of palladium as catalyst with phosphinite-ligands immobilized on silica and hexagonal mesoporous silica (HMS) having palladium nanoparticles as catalyst for Heck coupling reaction (HCR). Mono-, di-, tri-phosphinite ligands were functionalized on silica and hydroxyl functionalized HMS. Then, phosphinite functionalized silica/HMS-based palladium complexes were formed. The recycling ability of the catalyst was also determined and reported. Higher efficiency and stability were shown by Pd catalyst for HCR, whereas HMS-based catalyst show less efficiency and this was attributed to the time taking process of insertion of substrate into pores and thus renders onto its efficiency as catalyst. Chen et al. [53] reported new nucleophilic catalytic system based on dialkylaminopyridine functionalized mesoporous silica nanospheres (DMAP-MSNs). Initially, DMAP was functionalized with triethoxysilyl groups to produce 4-[*N*-[3-(triethoxysilyl)propyl]-*N*-methylamino]pyridine (DMAP-TES). DMAP-TES were then functionalized into MSNs via co-condensation to produce DMAP-MSNs. Finally, these DMAP-MSNs were utilized for Baylis-Hillman, acylation, and silylation catalytic reactions. It was observed that by this approach, the protonation of DMAP was avoided, which produced a heterogeneous catalyst with high reactivity and recyclability for Baylis-Hillman, acylation, and silylation reactions.

An upcoming research area of MSNs is to utilize it as an efficient material for the generation of clean energy, for example, hydrogen. It is a major challenge when it comes for hydrogen storage and release application. Different microporous materials particularly metal organic frameworks (MOFs) have made an impact on hydrogen storage and release applications. MSNs have created a significant interest to work on due to its large mesoporous channels that allow easy storage of chemical hydrides. Peru et al. [54] reported nanoconfinement of sodium borohydride ($NaBH_4$) into the mesoporous architecture of highly ordered SBA-15 and carbon-based materials (C-based CMK3) by using embedding approach such as wet chemical impregnation which is solvent (ammonia) free route. By such nanoconfinement process, there exists many advantages, for example, particle size of the activated phases can be precisely modulated, depending upon the nanoscale materials used; thus, alterations can be performed in the properties of nanoconfined particles. Also, major benefits of having nanoconfinement are not only reversibility observed for absorption/desorption cycle of hydrogen gas but also safe handling of the system. Thus, nanosizing leads to low temperature (below 120°C) desorption for $NaBH_4$ dispersed in mesoporous scaffolds of silica as well as carbon-based systems. Similarly, Lai et al. [55] reported the thermal behavior of ammonia borane (AB) embedded in mesoporous scaffolds of silica-based nanoparticles (SBA-15 and MCM-41). It was observed that dehydrogenation temperature for hydrogen release was lowered and the liberation of borazine and diborane as side products was minimized. The ammonia borane incorporated in MCM-41 showed favorable results than SBA-15. The was possibly due to pore size difference in MCM-41 and SBA-15. Smaller particles of AB molecules are more easily formed in MCM-41 having higher surface area and smaller pore size than SBA-15 leading to intensive contact of these particles onto the scaffolds of MCM-41 than SBA-15. The formation of coordination bond as O: → BH_3 occurs which reduces the H-bonding (intermolecular) in AB particles; thus, the release of hydrogen was lowered and simultaneously strong bonding of BH_3 with scaffolds remains intact and reduces diborane and borazine release. Deka et al. [56] reported confinement and fabrication of Ru nanoparticles in S1B-C10 (cubic MSNs modified via carboxylic acid functionality) using double chemical reduction approach. The 3D cage structure of mesoporous silica SBA-1 offers direct and effective growth of nanoparticles and also avoids agglomerations of these nanoparticles. The carboxylic acid functionality within the mesoporous silica ensures the great dispersion and size control of Ru metal NPs and thus limits its escape from the recyclable system and also enhances the stability of the catalyst. The ordered porous structure of MSNs can be utilized for preparation of shear thickening fluids (STF). Such behavior of MSNs finds application in flexible protective gear. It has been demonstrated that when MSNs are dispersed in any polar solvent/fluid, its high surface area and rough surface would enhance the interfacial interaction between the MSNs and dispersing medium, resulting in shear thickening behavior, in which an increase in viscosity has been observed with an increase in the shear rate. He et al. [57] studied the shear thickening behavior of suspensions prepared using porous silica nanoparticles. It was observed that the suspension remarked STF behavior with 42.5 wt.% concentration of nanoparticles which is much lower than the STF of non-porous silica nanoparticles. The possible mechanism for such behavior due to interfacial interactions was studied and was successfully demonstrated that it can be used in engineering composite applications.

4. Conclusions

MSNs are an important substrate possessing several favorable characteristics, for example, tunable pore size, surface area, and possibility of functionalization using wide variety of strategies. These important features makes MSNs widely applicable in the biomedical area, as sensors and catalysts. The grafting of a stimuli-responsive polymer from MSNs makes then suitable for drug-delivery application. We have summarized in this chapter, various controlled radical polymerization techniques, for example, RAFT, ATRP, and NMP for grafting of polymer chains. There is still a lot of scope in the synthesis and characterization of MSNs via co-condensation approach. In addition, in-depth characterization and mechanism of synthesis of MSNs in the presence of organoalkoxysilane can help in understanding the preferential location of organoalkoxysilane moieties after co-condensation. The modified/functionalized MSNs can also find application as a source of clean energy and novel composites with application in the area of protective materials.

Author details

Leena Nebhani*, Smrutirekha Mishra and Tina Joshi
Department of Materials Science and Engineering, Indian Institute of Technology Delhi, New Delhi, India

*Address all correspondence to: leena.nebhani@mse.iitd.ac.in

References

[1] Beck JS, Vartuli JC, Roth WJ, Leonowicz ME, Kresge CT, Schmitt KD, et al. A new family of mesoporous molecular sieves prepared with liquid crystal templates. Journal of the American Chemical Society. 1992;**114**:10834-10843

[2] Barton TJ, Bull LM, Klemperer WG, Loy DA, McEnaney B, Misono M, et al. Tailored porous materials. Chemistry of Materials. 1999;**11**:2633-2656

[3] Danks AE, Hall SR, Schnepp Z. The evolution of "sol-gel" chemistry as a technique for materials synthesis. Materials Horizons. 2016;**3**:91-112

[4] Alothman ZA. A review: Fundamental aspects of silicate mesoporous materials. Materials (Basel). 2012;**5**:2874-2902

[5] Zhao XS, Lu GQ, Millar GJ. Advances in mesoporous molecular sieve MCM-41. Industrial and Engineering Chemistry Research. 1996;**35**:2075-2090

[6] Kruk M, Jaroniec M, Ryoo R, Kim JM. Characterization of high-quality MCM-48 and SBA-1 mesoporous silicas. Chemistry of Materials. 1999;**11**:2568-2572

[7] Che S, Sakamoto Y, Terasaki O, Tatsumi T. Control of crystal morphology of SBA-1 mesoporous silica. Chemistry of Materials. 2001;**13**:2237-2239

[8] Javad Kalbasi R, Zirakbash A. Synthesis, characterization and drug release studies of poly(2-hydroxyethyl methacrylate)/KIT-5 nanocomposite as an innovative organic-inorganic hybrid carrier system. RSC Advances. 2015;**5**:12463-12471

[9] Cao L, Kruk M. Grafting of polymer brushes from nanopore surface via atom transfer radical polymerization with activators regenerated by electron transfer. Polymer Chemistry. 2010;**1**:97-101

[10] Raman NK, Anderson MT, Brinker CJ. Template-based approaches to the preparation of amorphous, nanoporous silicas. Chemistry of Materials. 1996;**8**:1682-1701

[11] Pelegri-O'Day EM, Maynard HD. Controlled radical polymerization as an enabling approach for the next generation of protein-polymer conjugates. Accounts of Chemical Research. 2016;**49**:1777-1785

[12] Möller K, Bein T. Talented mesoporous silica nanoparticles. Chemistry of Materials. 2017;**29**:371-388

[13] Huo Q, Margolese DI, Stucky GD. Surfactant control of phases in the synthesis of mesoporous silica-based materials. Chemistry of Materials. 1996;**8**:1147-1160

[14] Björk EM. Synthesizing and characterizing mesoporous silica SBA-15: A hands-on laboratory experiment for undergraduates using various instrumental techniques. Journal of Chemical Education. 2017;**94**:91-94

[15] Kruk M, Jaroniec M, Ko CH, Ryoo R. Characterization of the porous structure of SBA-15. Chemistry of Materials. 2000;**12**:1961-1968

[16] Feliczak-Guzik A, Jadach B, Piotrowska H, Murias M, Lulek J, Nowak I. Synthesis and characterization of SBA-16 type mesoporous materials containing amine groups. Microporous and Mesoporous Materials. 2016;**220**:231-238

[17] Yamamoto T, Mori S, Kawaguchi T, Tanaka T, Nakanishi K, Ohta T, et al. Evidence of a strained pore wall structure in mesoporous silica FSM-16 studied by X-ray absorption spectroscopy. Journal of Physical Chemistry C. 2008;**112**:328-331

[18] Heikkilä T, Salonen J, Tuura J, Hamdy MS, Mul G, Kumar N, et al. Mesoporous silica material TUD-1 as a drug delivery system. International Journal of Pharmaceutics. 2007;**331**: 133-138

[19] Barrabino A. Synthesis of Mesoporous Silica Particles with Control of Both Pore Diameter and Particle Size. PhD Thesis. Göteborg, Sweden: Chalmers university of technology; 2011. pp. 1-63

[20] Jammaer J, Aerts A, D'Haen J, Seo JW, Martens JA. Synthesis and characterization of COK-12 ordered mesoporous silica at room temperature under buffered quasi neutral pH. Studies in Surface Science and Catalysis. 2010;**175**;681-684

[21] Kruk M, Jaroniec M, Kim JM, Ryoo R. Characterization of highly ordered MCM-41 silicas using X-ray diffraction and nitrogen adsorption. Langmuir. 1999;**15**:5279-5284. DOI: 10.1021/la990179v

[22] Li ZH, Gong YJ, Wu D, Sun YH, Wang J, Liu Y, et al. SAXS analysis of interface in organo-modified mesoporous silica. Surface and Interface Analysis. 2001;**31**:897-900

[23] Trébosc J, Wiench JW, Huh S, Lin VSY, Pruski M. Solid-state MMR study of MCM-41-type mesoporous silica nanoparticles. Journal of the American Chemical Society. 2005;**127**:3057-3068

[24] Kruk M, Jaroniec M, Sakamoto Y, Terasaki O, Ryoo R, Ko CH. Determination of pore size and pore wall structure of MCM-41 by using nitrogen adsorption, transmission electron microscopy, and X-ray diffraction. The Journal of Physical Chemistry. B. 2000;**104**:292-301

[25] Li T, Senesi AJ, Lee B. Small angle X-ray scattering for nanoparticle research. Chemical Reviews. 2016;**116**:11128-11180

[26] Yi Z, Dumée LF, Garvey CJ, Feng C, She F, Rookes JE, et al. A new insight into growth mechanism and kinetics of mesoporous silica nanoparticles by in situ small angle X-ray scattering. Langmuir. 2015;**31**:8478-8487

[27] Zhao D, Sun J, Li Q, Stucky GD. Morphological control of highly ordered mesoporous silica SBA-15. Chemistry of Materials. 2000;**12**:275-279

[28] Barbey R, Lavanant L, Paripovic D, Schüwer N, Sugnaux C, Tugulu S, et al. Polymer brushes via surface-initiated controlled radical polymerization: Synthesis, characterization, properties, and applications. Chemical Reviews. 2009;**109**:5437-5527

[29] Chiefari J, Chong YK, Ercole F, Krstina J, Jeffery J, Le TPT, et al. Living free-radical polymerization by reversible addition—Fragmentation chain transfer: The RAFT process. Macromolecules. 1998;**31**:5559-5562

[30] Ma M, Zheng S, Chen H, Yao M, Zhang K, Jia X, et al. A combined "rAFT" and "graft from" polymerization strategy for surface modification of mesoporous silica nanoparticles: Towards enhanced tumor accumulation and cancer therapy efficacy. Journal of Materials Chemistry B. 2014;**2**:5828-5836

[31] Mishra S, Rawal A, Nebhani L. Imprinting the location of an in-built RAFT agent and selective grafting of polymer chains inside or outside the pores of mesoporous silica nanoparticles. Microporous and Mesoporous Materials. 2020;**294**:109898

[32] Mishra S, Hook JM, Nebhani L. Priming the pores of mesoporous silica nanoparticles with an in-built RAFT agent for anchoring a thermally

responsive polymer. Microporous and Mesoporous Materials. 2019;**277**:60-69

[33] Hong CY, Li X, Pan CY. Grafting polymer nanoshell onto the exterior surface of mesoporous silica nanoparticles via surface reversible addition-fragmentation chain transfer polymerization. European Polymer Journal. 2007;**43**:4114-4122

[34] Gonçalves JLM, Crucho CIC, Alves SPC, Baleizão C, Farinha JPS. Hybrid mesoporous nanoparticles for pH-actuated controlled release. Nanomaterials. 2019;**9**:1-13

[35] Huang L, Liu M, Mao L, Xu D, Wan Q, Zeng G, et al. Preparation and controlled drug delivery applications of mesoporous silica polymer nanocomposites through the visible light induced surface-initiated ATRP. Applied Surface Science. 2017;**412**:571-577

[36] Zhou Z, Zhu S, Zhang D. Grafting of thermo-responsive polymer inside mesoporous silica with large pore size using ATRP and investigation of its use in drug release. Journal of Materials Chemistry. 2007;**17**:2428

[37] Zhou H, Wang X, Tang J, Yang YW. Tuning the growth, crosslinking, and gating effect of disulfide-containing PGMAs on the surfaces of mesoporous silica nanoparticles for redox/pH dual-controlled cargo release. Polymer Chemistry. 2016;7:2171-2179

[38] Gigmes D. Nitroxide mediated polymerization: From fundamental to applications in materials science. Royal Society of Chemistry. 2015;**19**;1-120

[39] Charleux B. Surface-initiated nitroxide-mediated polymerization from ordered mesoporous silica. Macromolecules. 2011;**44**(8):2577-2588

[40] Bartholome C, Beyou E, Bourgeat-Lami E, Chaumont P, Lefebvre F, Zydowicz N. Nitroxide-mediated polymerization of styrene initiated from the surface of silica nanoparticles: In situ generation and grafting of alkoxyamine initiators. Macromolecules. 2005;**38**:1099-1106

[41] Sahoo S, Bordoloi A, Halligudi SB. Ordered mesoporous silica as supports in the heterogeneous asymmetric catalysis. Catalysis Surveys from Asia. 2011;**15**:200-214

[42] Möller K, Bein T. Degradable drug carriers: Vanishing mesoporous silica nanoparticles. Chemistry of Materials. 2019;**31**:4364-4378

[43] Guillet-Nicolas R, Marcoux L, Kleitz F. Insights into pore surface modification of mesoporous polymer-silica composites: Introduction of reactive amines. New Journal of Chemistry. 2010;**34**:355-366

[44] Tarn D, Ashley CE, Xue M, Carnes EC, Zink JI, Brinker CJ. Mesoporous silica nanoparticle nanocarriers: Biofunctionality and biocompatibility. Accounts of Chemical Research. 2013;**46**:792-801

[45] Paris JL, Cabanas MV, Manzano M, Vallet-Regí M. Polymer-grafted mesoporous silica nanoparticles as ultrasound-responsive drug carriers. ACS Nano. 2015;**9**:11023-11033

[46] Zheng Y, Wang L, Lu L, Wang Q, Benicewicz BC. pH and Thermal and dual-responsive nanoparticles for controlled drug delivery with high loading content. ACS Omega. 2017;**2**:3399-3405

[47] Liu X, Yu D, Jin C, Song X, Cheng J, Zhao X, et al. A dual responsive targeted drug delivery system based on smart polymer coated mesoporous silica for laryngeal carcinoma treatment. New Journal of Chemistry. 2014;**38**:4830-4836

[48] Tang D, Zhang L, Zhang Y, Qiao ZA, Liu Y, Huo Q. Mesoporous

silica nanoparticles immobilized salicylaldimine cobalt complexes as high efficient catalysts for polymerization of 1,3-butadiene. Journal of Colloid and Interface Science. 2012;**369**:338-343

[49] Tang D, Zhang W, Qiao Z, Liu Y, Huo Q. Functionalized mesoporous silica nanoparticles as a catalyst to synthesize a luminescent polymer/silica nanocomposite. RSC Advances. 2016;**6**:16461-16466

[50] Albuquerque MCG, Jiménez-Urbistondo I, Santamaría-González J, Mérida-Robles JM, Moreno-Tost R, Rodríguez-Castellón E, et al. CaO supported on mesoporous silicas as basic catalysts for transesterification reactions. Applied Catalysis A: General. 2008;**334**:35-43

[51] Abdullah AZ, Razali N, Lee KT. Optimization of mesoporous K/SBA-15 catalyzed transesterification of palm oil using response surface methodology. Fuel Processing Technology. 2009;**90**:958-964

[52] Farjadian F, Hosseini M, Ghasemi S, Tamami B. Phosphinite-functionalized silica and hexagonal mesoporous silica containing palladium nanoparticles in Heck coupling reaction: Synthesis, characterization, and catalytic activity. RSC Advances. 2015;**5**:79976-79987

[53] Chen HT, Huh S, Wiench JW, Pruski M, Lin VSY. Dialkylamino-pyridine-functionalized mesoporous silica nanosphere as an efficient and highly stable heterogeneous nucleophilic catalyst. Journal of the American Chemical Society. 2005;**127**:13305-13311

[54] Peru F, Garroni S, Campesi R, Milanese C, Marini A, Pellicer E, et al. Ammonia-free infiltration of NaBH4 into highly-ordered mesoporous silica and carbon matrices for hydrogen storage. Journal of Alloys and Compounds. 2013;**580**:2-5

[55] Lai SW, Lin HL, Yu TL, Lee LP, Weng BJ. Hydrogen release from ammonia borane embedded in mesoporous silica scaffolds: SBA-15 and MCM-41. International Journal of Hydrogen Energy. 2012;**37**:14393-14404

[56] Deka JR, Saikia D, Hsia K-S, Kao H-M, Yang Y-C, Chen C-S. Ru nanoparticles embedded in cubic mesoporous silica SBA-1 as highly efficient catalysts for hydrogen generation from ammonia borane. Catalysts. 2020;**10**:267

[57] He Q, Gong X, Xuan S, Jiang W, Chen Q. Shear thickening of suspensions of porous silica nanoparticles. Journal of Materials Science. 2015;**50**:6041-6049

Chapter 5

Melting and Dissolving Fly Ash by NaOH for the Removal of Iron, Calcium, and Other Impurities

Minghua Wang, Hui Zhao, Gulambar Tursun, Jianwang Yang and Long Liu

Abstract

A mixture of fly ash and sodium hydroxide was calcined, which converted mullite ($3Al_2O_3{\cdot}2SiO_2$), the high-temperature stability phase containing silicon and aluminum oxides, and quartz into activated silica alumina phase, and they were dissolved by concentrated NaOH solution into soluble SiO_3^{2-} and AlO_2^{-}. The insoluble impurities including Fe_2O_3, FeO, CaO, and $CaSO_4$ were filtered out. Experiment results show the optimum experimental conditions for the dissolution: temperature is 60°C, 15% NaOH solution is used, liquid–solid mass ratio is 11:1, stirring time is 3 h, and about 78.9% of silicon and 78.1% of aluminum are dissolved. The obtained pure silicon aluminum solution provides the raw material for preparing high-purity molecular sieves, and Fe_2O_3 content of the prepared P-type molecular sieve is only 0.25%, and the CaO content is only 0.066%. The paper provided a viable method to remove Fe, Ca, etc. in mineral thoroughly.

Keywords: melting and dissolving of coal fly ash, removal of Fe, P molecular sieve

1. Introduction

Fly ash is a kind of solid waste discharged from power plants and various coal-fired boilers. According to statistics, the amount of fly ash accumulation in China is up to 0.2 billion tons, and it is increasing every year. It is the largest output of ash in industrial waste residue. Fly ash can seriously pollute the environment and cause serious harm to people's lives, animals and plants, and so on. According to local conditions, timely and effective treatment of fly ash and comprehensive utilization of fly ash have far-reaching significance. It not only saves water, saves soil, and turns waste into treasure but also protects the environment.

At present, fly ash is mainly used as building material. It is used in thermal insulation board, slag cement, wall tile, floor brick, etc. [1]. The utilization ratio of fly ash is large, reaching 67%, but the added value of products is not high. The main representative components of fly ash are silica and alumina, which account for as high as 70%, while the silicon and aluminum are also main ingredients of expensive molecular sieve. Preparing high price molecular sieve with cheap fly ash is an important way to promote the use of additional value. The application of molecular sieve is very wide; can be used for gas or liquid dehydration, drying, separation, and purification; and can also be used as adsorbent, catalyst, and ion exchange agent for various types of reactions in the field of petroleum chemical industry, fine chemical industry, agriculture, and environmental protection. The preparation of molecular sieve with fly ash can not only save raw materials but also can simplify the process and equipment and provide the conditions for large-scale production and wide application of molecular sieves.

The study on the synthesis of molecular sieves with fly ash began from 1985 by Holler and Wrisching [2].

Since then, more and more molecular sieve types [3] have been developed. The domestic and international research on the synthesis of molecular sieve from fly ash is more and more widely and deeply [4].

Yang Liyun et al. recently synthesized zeolite 4A using fly ash fused with synergism of NaOH and Na_2CO_3 [5]. Asifa Iqbal et al. synthesized and characterized pure phase 4A from coal fly ash [6]. Richa Soni and Dericks Praise Shukia synthesized fly ash-based zeolite-reduced graphene oxide composite and evaluated its property as an adsorbent for arsenic removal [7]. A De Rossi et al. synthesized zeolite by geopolymerization of biomass fly ash and metakaolin [8]. Koshy and Singh described applications of fly ash zeolites for water treatment [9]. Fang et al. synthesized high-quality zeolites from coal fly ash and researched mobility of hazardous elements and environmental applications [10]. Tauanov et al. synthesized coal fly ash-derived zeolites doped with silver nanoparticles for mercury(II) removal from water [11]. Lim et al. prepared quasi-solid-state electrolytes using a coal fly ash-derived zeolite-X and zeolite-A for dye-sensitized solar cells [12]. Collectively, researches on zeolites derived from fly ash focus on applications gradually.

However, fly ash is not pure aluminum silicate. In addition to containing valuable elements of silicon and aluminum, fly ash also contains a considerable part of the iron and calcium and other impurities, leading to impure molecular sieve. The impurities not only are easy to plug in the channels of the molecular sieve but also reduce the exchange capacity, catalytic performance, and cycle performance. The drawbacks have not drawn enough attentions so far. Therefore, it is necessary to pretreat fly ash to remove the impurities such as iron, calcium, and so on before preparing the molecular sieve with fly ash.

Acid leaching of fly ash to remove iron and calcium used to be employed by previous patent literature, as the reaction temperature is below 100°C, the process can not destroy the high-temperature phase such as mullite and quartz, so the high-temperature phase included iron and calcium impurities does not dissolve, this process can only get rid of 60% of iron and calcium, is not complete. We once used carbon reduction-magnetic separation-acid leaching method; although the iron removal efficiency is high, the process is slightly lengthy [13]. Alkali melting method was used in literature, but the following alkali dissolving was not used. The impurities still exist in prepared molecular sieve [14].

A new method, alkali melting, or alkali dissolving, of fly ash, to remove iron, calcium, and other impurities, was proposed in the paper. Firstly, fly ash and NaOH solid were mixed and roasted to convert the mullite and quartz into glass phases. The roasted clinker was leached by NaOH solution to dissolve silicon and aluminum components and to filter out Fe_2O_3, CaO, and $CaSO_4$ insoluble impurities, etc., so as to obtain purified Na_2SiO_3 and $NaAlO_2$ solution. The purified solution can be used for the preparation of high pure molecular sieve to improve the performance and the service life.

2. Experimental part

In this study, the fly ash was gotten from a power plant in Shandong; the main chemical composition is shown in **Table 1**.

Ingredient	SiO_2	Al_2O_3	$CaSO_4$	MgO	Na_2O	Fe_2O_3	Others
Percent (%)	41.36	30.45	24.75	1.15	0.26	3.70	1.66

Table 1.
Representative components of the fly ash (mass fraction, %).

The specific steps are as follows. A certain amount of fly ash was weighed, placed in a mortar, and fully ground before putting through a 200 mesh sieve prior to be mixed with NaOH powder with a mass ratio of 1 to 2. The mixture was transferred to a crucible and was roasted at 600°C in a muffle furnace for 2 h. The clinker was then leached with a high concentration of NaOH solution to dissolve the silicon aluminum phase, and the impurities such as iron and calcium are filtered out to obtain a pure solution containing the silicon aluminum phase merely. The contents of Si and Al in the filtrate were analyzed by ICP analyzer, and the dissolution ratios of silicon and aluminum were calculated according to the total amounts. At the same time, the dissolution ratio of fly ash in clinker was also measured. The filtrate was added with hydrochloric acid and the alkalinity (OH/Si) was controlled. A certain proportion of Na_2SiO_3 solution and the seed crystal directing agent was added to regulate the ratio of silicon and aluminum for a special molecular type. After that, the mixture was stirred evenly, transferred to an autoclave, and crystallized at a certain temperature for a certain time. The derived solid was just a molecular sieve and was filtered and dried, measured by SHIMADZU X-ray diffractometer XRD-6000. Zeolite P was synthesized according to the document [14]. Ion concentration is analyzed by 3600A inductively coupled plasma atomic emission spectrometer made by Keje company in China.

3. Results and discussion

3.1 Calcination of NaOH and fly ash

SiO_2 and Al_2O_3 existing as quartz and mullite in fly ash have low activity; high-temperature alkali roasting method can greatly increase their activity and improve the efficiency of the fly ash conversion and crystallization synthesis of the molecular sieve. There are commonly two ways to calcine fly ash, one is to add Na_2CO_3 to calcine at 800°C, and the other is to add NaOH to calcine at 600°C. In the experiment, fly ash and NaOH reacted after roasted in a muffle furnace at 600°C for 2 h. **Figure 1** is the contrast diagram of XRD for fly ash before and after calcination with NaOH.

From **Figure 1**, the diffraction peaks of quartz and mullite are strong before calcination, which show the main minerals are quartz and mullite. Quartz and mullite are difficult to react with NaOH at room temperature due to their low activity. Therefore, calcination of fly ash and NaOH at high temperature is necessary and can stimulate its activity. The diffraction pattern of calcined fly ash and NaOH shows that the form of the material is mainly aluminosilicate when calcined at high temperature. At this time, there are few quartz and mullite, and the diffraction peak almost disappeared. This is because as the reaction proceeded, high-temperature roasting destroyed crystal structure, thus releasing the active SiO_2 and Al_2O_3. These substances react with NaOH and generate amorphous aluminosilicate that is able to participate in the zeolite framework structure.

$$3Al_2O_3 \cdot 2SiO_2 + NaOH \rightarrow NaAlO_2 + Na_2SiO_3 + H_2O \tag{1}$$

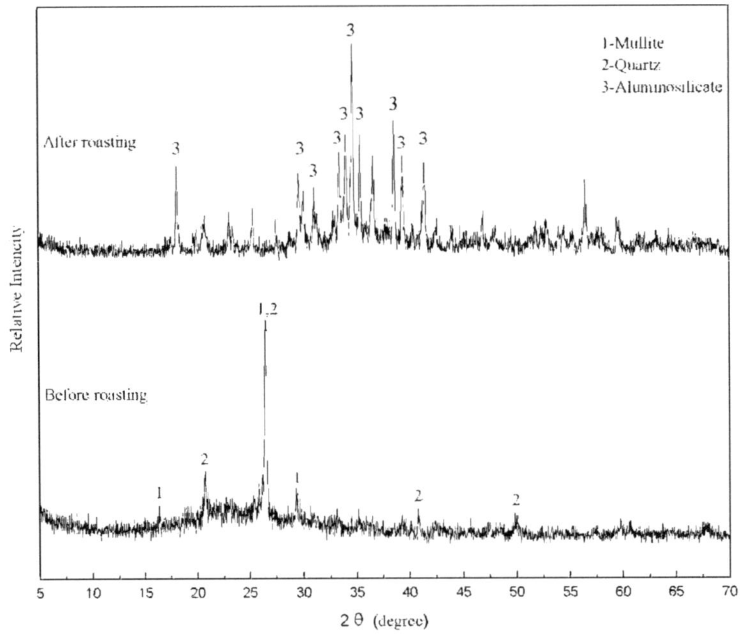

Figure 1.
Effect of alkali melting activation on fly ash structure.

$$SiO_2[quartz] \rightarrow SiO_2[glass] \quad (2)$$

What is more, high-temperature roasting can get rid of the organic impurities in fly ash and amorphous carbon, thereby improving the purity of raw materials. In addition, alkali melting of fly ash provided a large amount of NaOH for the following leaching process since real mass amount of NaOH reacted is equal to the of fly ash according to the experimental result.

3.2 Effect of mass ratio of solid clinker to alkali solution

Figure 2 shows that the NaOH solution can dissolve alkali melting residue, which includes Na_2SiO_3 and $NaAl(OH)_4$ derived from mullite, quartz, and other silicon aluminum phases of coal fly ash. Dissolving ratios of Si and Al decrease with increasing mass ratio of solid clinker to alkali solution. The maximum dissolving ratio of Si is 56.6%, while the maximum dissolving ratio of Al is 33.8% under the experimental condition. 100 g of water can dissolve nearly 37 g Na_2SiO_3; however, 56.6% is not high.

3.3 Effect of temperature on dissolving ratio of Si and Al

From **Figure 3**, low leaching temperature is beneficial to high dissolving ratio of Si and Al. Dissolving ratios of Si and Al reach 78.9 and 78.1%, respectively, as the leaching temperature is 20°C. Similar research showed dissolution ratio of Si is 75%, while dissolution ratio of Al is only 25% during alkali leaching from titania slag [15]. Both the dissolving ratios dropped sharply with increasing leaching temperatures because hydrolysis reaction occurred at higher temperatures. H_4SiO_4 and $Al(OH)_3$ precipitates stayed in solid residues, which led to the low dissolving ratios.

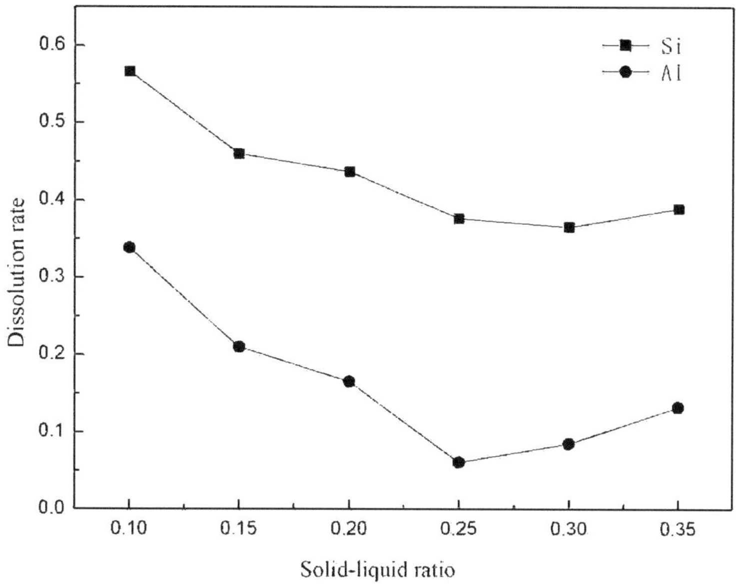

Figure 2.
Effect of mass ratio of solid clinker to alkali solution 15% NaOH, 1 h of stirring time, 60°C.

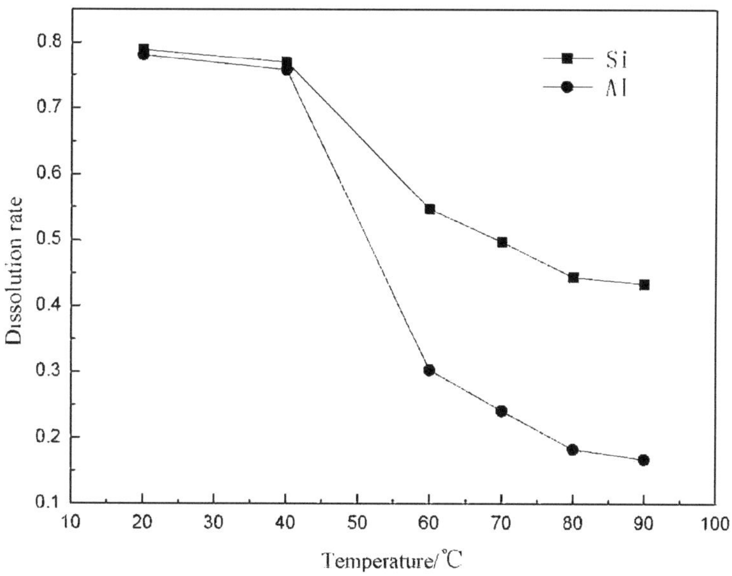

Figure 3.
Effect of temperature on dissolving ratio of Si and Al 15% NaOH, NaOH solution clinker mass ratio of 10:1, 1 h of stirring time.

3.4 Effect of mass ratio of NaOH to solution on dissolving ratio of Si and Al

Figure 4 shows higher NaOH concentration is more viable in order to dissolve more Si and Al. When the alkali concentration is increased from 5 to 10%, the dissolution rate of aluminum decreases. The leaching rate fluctuated, just like the acid leaching of titanium-bearing minerals; at present it is thought that the evolution of AlO_2^- causes the fluctuation because the AlO_2^- is more soluble than $Al(OH)_4^-$, and the general trend is still rising with the increase of NaOH concentration.

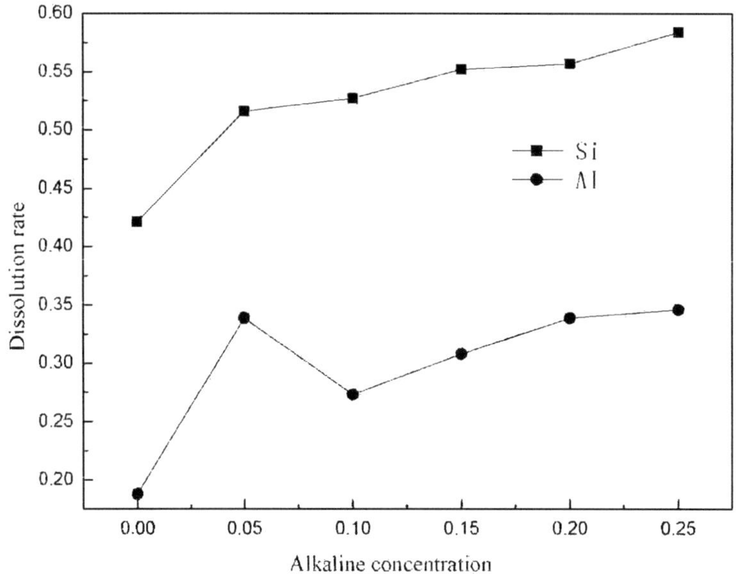

Figure 4.
Effect of mass ratio of NaOH to solution on dissolving ratio of Si and Al 60°C, NaOH solution to clinker mass ratio of 10:1, 1 h of stirring time.

The possible following reaction occurs when NaOH concentration is larger than 10%.

$$Na_2O \cdot Al_2O_3 \cdot xSiO_2 + 2NaOH + 4H_2O \rightarrow 2NaAl(OH)_4 + Na_2[H_2SiO_4]$$

Thus, insoluble $Na_2O \cdot Al_2O_3 \cdot xSiO_2$ in water dissolved in concentrated NaOH solution. The tendency indicates OH- plays an important role during the process, which takes part in the coordination process. The large amount of use of NaOH causes big cost. Reusing the redundant NaOH solution is significant and economical. The condensation and the seed can make the reuse of NaOH solution possible.

Figure 5 shows Na_2O-Al_2O_3-H_2O system phase diagram [16]. The area of OBCO belongs to the dissolving process of Al_2O_3 into the NaOH solution according to **Figure 3**. The solubility of Al_2O_3 in NaOH solution is increasing along the NaOH concentration until 20% (line OB); after the summit of 20% of the solubility, the dissolving ability decreases with the increasing NaOH concentration [line BC]. Therefore, a highly efficient NaOH concentration is around 20%; the number is consistent with our experimental result in **Figure 4** where 0.20 mass ratio of NaOH made the highest dissolving ratio of Al. 15% NaOH is used, and NaOH is surplus after the roasting process. Overall, the concentration of NaOH is around 20%.

3.5 Effect of stirring time on dissolving ratios of Si and Al

Figure 6 indicates too long time is disadvantageous for high dissolving ratio of Si and Al. The dissolving ratio decreases with adding stirring time. The tendency is obvious for Al. The dissolving ratio drops steeply during 2 h or so. The gradual decreasing tendency is due to the precipitate of Si and Al. Reducing stirring time is essential so as to get high dissolving ratios of Si and Al. 1 h is enough for the stirring time.

Figure 7 shows the filtration of slag after alkaline solution filtration. The color of the slag is red, which proves that the slag contains iron oxide, which can be used for carbon reduction to obtain sponge iron for steel-making. The filtered solution can

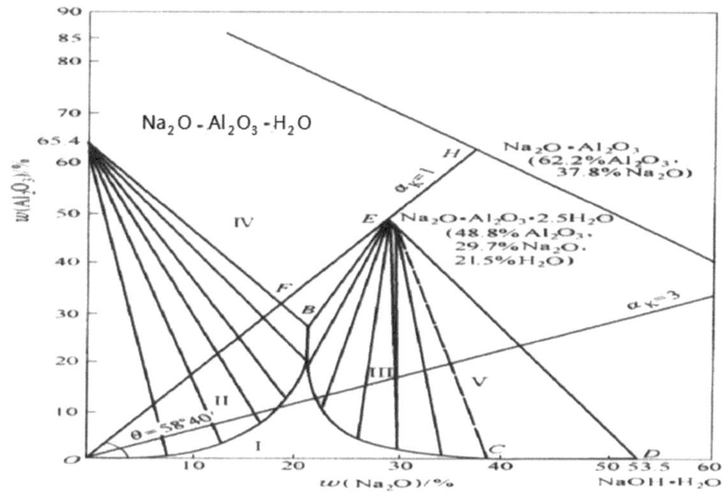

Figure 5.
Na_2O-Al_2O_3-H_2O phase diagram.

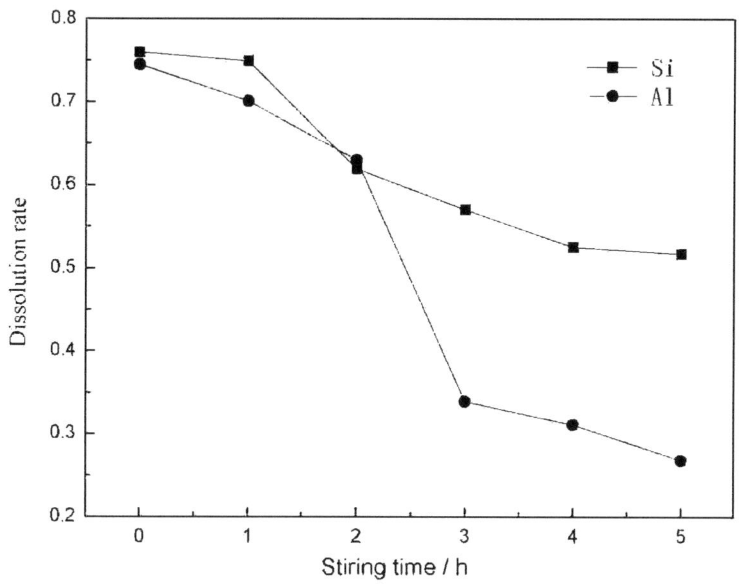

Figure 6.
Effect of stirring time on dissolving ratios of Si and Al 60°C, NaOH solution to clinker mass ratio of 10:1.

be used for the preparation of the molecular sieve. The molar ratio of Si to Al in the leach solution is adjusted to 4.6; pH drops to 9 by adding HCl solution. Then, the solution is put in an autoclave after aging for 1 h and taken out after crystallization for 16 h at 100°C. The p-type molecular sieve is obtained after filtration, washing, and drying. The cost of raw material and the environmental pollution were reduced because of using fly ash as raw material.

The Fe_2O_3 content of the prepared molecular sieve drops to 0.25% after the alkali treatment from 2.87% by merely acid leaching of fly ash, while the content of CaO drops to 0.066% from 1.18%. The chemical formula of the P molecular sieve prepared is confirmed as $Na_6Al_6Si_{10}O_{32}{\cdot}12H_2O$ through ICP examination. The purity of the P molecular sieve prepared by the method from coal fly ash

Figure 7.
Residue of insoluble clinker after filtration.

Ingredient	SiO_2	Al_2O_3	Na_2O	K_2O	Fe_2O_3	CaO	SO_3
Percent (%)	58.74	25.65	14.67	0.61	0.25	0.066	0.011

Table 2.
Chemical composition of P molecular sieve prepared by fly ash (mass fraction, %).

Figure 8.
P molecular sieve prepared by coal fly ash.

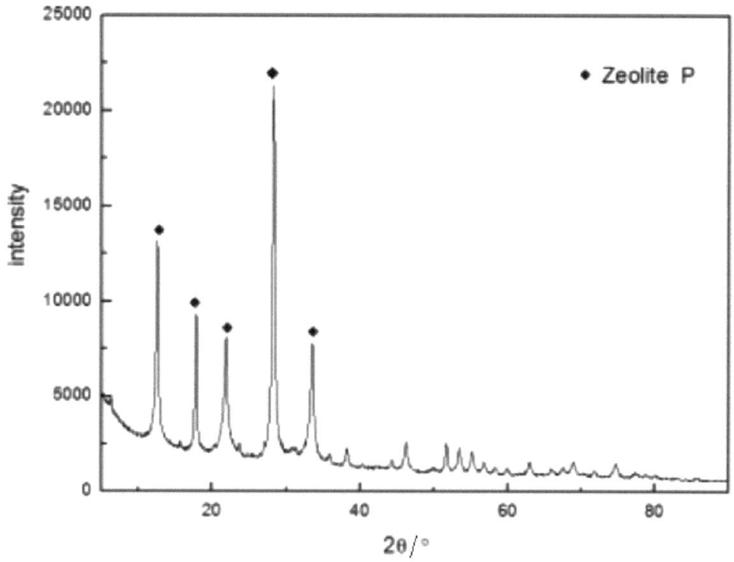

Figure 9.
XRD diagram of P molecular sieve prepared by coal fly ash.

is 99.06% (seen in **Table 2**), while the whiteness level is 96 (seen in **Figure 8**). Both purity and whiteness level of the molecular sieve can meet the standard of market molecular sieve.

Figure 9 shows XRD diagram of the product; there are five characteristic peaks at 2θ = 12.48°, 17.71°, 21.68°, 28.15°, and 33.37°, respectively, and miscellaneous peaks are few, indicating the product is P molecular sieve (*JCPDS* NO 39-0219). Pure P molecular sieve can be used as petroleum catalyst, support, and detergent additive to soften hard water in washing process.

4. Conclusion

At 600°C, the alkali melting activated quartz and mullite in the fly ash into the aluminosilicate glass phase, which can be dissolved in the alkali solution. The optimum conditions of alkali dissolving of clinker were obtained by optimizing experiments, that is, the reaction temperature is 20°C; using 15% NaOH solution, the liquid solid ratio is 10:1, and the stirring time is less than 1 h; the whole dissolution ratio of silicon reaches 78.9%, while that of Al reaches 78.1%. The method provides pure solution for preparing molecular sieves from fly ash. The Fe_2O_3 content of the prepared molecular sieve through actual verification drops to 0.25% after the alkali treatment from 2.87% by merely acid leaching of fly ash, while the content of CaO drops to 0.066% from 1.18%.

Author details

Minghua Wang*, Hui Zhao, Gulambar Tursun, Jianwang Yang and Long Liu
School of Metallurgy, Northeastern University, Shenyang, China

*Address all correspondence to: wangmh@smm.neu.edu.cn

References

[1] Wu M. Research progress on modification and application of fly ash. Guangzhou Chemical Industry. 2017;**45**(8):22-24

[2] Lin W, Li G, Wang A, Wang P. Research status of preparation of molecular sieves by fly ash. Anhui Chemical Industry. 2017;**43**(1):9-14

[3] Yan Z, Wang X, Li H, Zhang J. Synthesis and characterization 13X zeolite prepared from desilication solution of high alumina fly ash and coal gangue. Bulletin of Chinese Ceramic Society. 2016;**36**(3):778-784

[4] Fukasawa T, Horigome A, Tsu T, et al. Utilization of incineration fly ash from biomass powder plants for zeolite synthesis from coal fly ash by hydrothermal treatment. Fuel Processing Technology. 2017;**167**:92-98

[5] Yang L, Qian X, Yuan P, Bao B, et al. Green synthesis of zeolite 4A using fly ash fused with synergism of NaOH and Na_2CO_3. Journal of Cleaner Production. 2019;**212**:250-260

[6] Iqbal A, Scattar H, Haider R, et al. Synthesis and characterization of pure phase zeolite 4A from coal fly ash. Journal of Cleaner Production. 2019;**219**:258-267

[7] Soni R, Shukia DP. Synthesis of fly ash based zeolite-reduced graphene oxide composite and its evaluation as an adsorbent for arsenic removal. Journal of Cleaner Production. 2019;**219**:504-509

[8] De Rossi A, Simao L, Ribeiro MJ, et al. In-situ synthesis of zeolites by geopolymerization of biomass fly ash and metakaolin. Materials Letters. 2019;**236**:644-648

[9] Koshy N, Singh DN. Fly ash zeolites for water treatment applications. Journal of Environmental Chemical Engineering. 2016;**4**(2):1460-1472

[10] Feng W, Wan Z, Daniels J, et al. Synthesis of high quality zeolites from coal fly ash: Mobility of hazardous elements and environmental applications. Journal of Cleaner Production. 2018;**202**:390-400

[11] Tauanov Z, Tsakiridis PE, Mikhaiosky SV, et al. Synthesized coal fly ash-derived zeolites doped with silver nanoparticles for mercury (II) removal from water. Journal of Environmental Management. 2018;**224**:164-171

[12] Lim J, M, Park JY, Park JT, et al. Prepared quasi-solid-state electrolytes using a coal fly ash derived zeolite-X and -A for dye-sensitized solar cells. Journal of Industrial and Engineering Chemistry. 2019;**71**:378-386

[13] Wang M, Zhao H, Liu Y, Kong C, Yang A, Li J. Removal of Fe from fly ash by carbon thermal reduction. Microporous & Mesoporous Materials. 2017;**245**:133-137

[14] Kong D, Song S, Wang Q, Lin S. Preparation of P-type molecular sieve from fly ash by alkali melting and hydrothermal method. Bulletin of Chinese Ceramic Society. 2016;**35**(3):922-926

[15] Lu H, Xie G, Yu X-H, et al. Investigation of silicon and aluminium removed mechanism by alkaline leaching from the titania slag in electric furnace. Light Metal. 2010;7:53-57

[16] Bi S. Production of Alumina by Bayer Process. Beijing: Metallurgical Industry Press; 2007. p. 7